高等职业院校新形态教材 · 大数据系列

Java Web 程序设计项目实战
（微课版）

主　编　廖　丽　范凌云
副主编　钱　游　王永峰　梁修荣

电子工业出版社
Publishing House of Electronics Industry
北京·BEIJING

内 容 简 介

本书根据高等职业院校的教学要求，以岗位需求为导向编写而成。本书结合真实的项目应用场景，将企业的"新技术、新工艺、新规范"有效融入教材中，系统阐述了 Java Web 相关知识，以帮助读者掌握岗位所必备的核心技能。

本书分为八个项目，包括 Java Web 开发环境准备、JSP 技术、JSP 内置对象、JDBC 与数据库访问、JavaBean 开发模型、Servlet 技术、EL 和 JSTL 技术、智慧金融信贷管理系统。各项目、各任务间层层递进，能帮助编程初学者快速上手、轻松入门，进而成长为编程高手。

本书附有配套课程标准、电子教案、电子课件、微视频、习题及答案等教学资源，读者可在书中扫描二维码获取，或登录华信教育资源网（www.hxedu.com.cn）免费下载。

本书可作为高等职业院校计算机相关专业的教材，也可作为广大计算机编程爱好者的自学教材。

未经许可，不得以任何方式复制或抄袭本书之部分或全部内容。
版权所有，侵权必究。

图书在版编目（CIP）数据

Java Web 程序设计项目实战：微课版/廖丽，范凌云主编. —北京：电子工业出版社，2023.8
ISBN 978-7-121-46210-8

Ⅰ.①J… Ⅱ.①廖… ②范… Ⅲ.①JAVA 语言—程序设计—高等职业教育—教材 Ⅳ.①TP312.8

中国国家版本馆 CIP 数据核字（2023）第 158324 号

责任编辑：魏建波
印　　刷：北京虎彩文化传播有限公司
装　　订：北京虎彩文化传播有限公司
出版发行：电子工业出版社
　　　　　北京市海淀区万寿路 173 信箱　邮编：100036
开　　本：787×1092　1/16　　印张：19.5　　字数：499.2 千字
版　　次：2023 年 8 月第 1 版
印　　次：2024 年 7 月第 2 次印刷
定　　价：59.00 元

凡所购买电子工业出版社图书有缺损问题，请向购买书店调换。若书店售缺，请与本社发行部联系，联系及邮购电话：（010）88254888，88258888。
质量投诉请发邮件至 zlts@phei.com.cn，盗版侵权举报请发邮件至 dbqq@phei.com.cn。
本书咨询联系方式：（010）88254173 或 qiurj@phei.com.cn。

前　　言

　　Java Web 是用 Java 技术来解决 Web 互联网相关领域的技术栈，其不仅拥有与 Java 一样的面向对象、跨平台、高度安全等优点和特性，还拥有 Java Servlet 的稳定性，并且可以应用 Servlet 提供的 API、JavaBean 及 Web 开发框架技术，使前端代码与后端代码分离，提高工作效率，因此，Java Web 成了大、中型网站开发的首选技术栈。

　　本书具有以下五个特色。

　　第一，采用"项目导入、任务促学"的设计思路。本书面向 Java 开发工程师的岗位需求，以职业岗位能力为依据，分析典型岗位需求，根据岗位的需求、学生的认知规律、高等职业教育"必需、够用"的理论教学原则来全面规划和重构教材内容。本书分为八个项目，包括 Java Web 开发环境准备、JSP 技术、JSP 内置对象、JDBC 与数据库访问、JavaBean 开发模型、Servlet 技术、EL 和 JSTL 技术、智慧金融信贷管理系统。各项目、各任务间层层递进，体现了"项目导入、任务促学"的设计思路。

　　第二，立足"教、学、做"一体化特色，设计"三位一体"的教材。从"教什么，怎么教""学什么，怎么学""做什么，怎么做"三个问题出发，并配套有课程标准、电子教案、电子课件、微视频、习题及答案等丰富的教学资源，同时将企业的"新技术、新工艺、新规范"有效融入教材中。

　　第三，遵循学生的认知规律和技能的掌握规律，按照"任务演示—知识准备—任务实施—任务拓展"的流程实施教学。通过任务演示让学生明确本次任务要达到的目标，围绕这个目标进行知识准备，再通过知识准备来实施任务，进而拓展提升，帮助编程初学者快速上手、轻松入门，进而成长为编程高手。

　　第四，有效整合教材内容与教学资源，打造立体化、自主学习式的新形态一体化教材。通过与项目任务对应的微视频二维码，让读者减少因为无教师引导实践操作而产生的学习难度。通过互联网（尤其是移动互联网）将数字教学资源与纸质教材融合，实现"新旧媒体融合"，成为"互联网+"时代教材功能升级和形式创新的成果。

　　第五，以真实项目为载体组织教学内容，实施教学过程，强化技能训练，提升动手能力。全书围绕"智慧金融信贷管理系统"综合项目所必备的技能点展开，采用任务驱动的教学方法，全方位促进学生对 Java Web 相关知识的理解，引导学生将知识点应用到具体实践中，并能够举一反三，满足就业岗位的要求。

　　本书的完成得益于重庆城市职业学院、科大讯飞股份有限公司校企混编团队，他们具备丰富的教学经验、项目开发实践经验，以及将理论知识转化为实际开发内容的能力。本书由廖丽、范凌云任主编，由钱游、王永峰、梁修荣任副主编。其中，项目一由梁修荣编写，项目二、项目三、项目五、项目七由廖丽编写，项目四由钱游编写，项目六由范凌云编写，项目八由王永峰编写。全书由廖丽统稿并审核。

本书配套完整的教学资源，请有需要的读者登录华信教育资源网（www.hxedu.com.cn）注册后免费下载相关配套资源，或扫描书中二维码获取。

资源名称	资源类别	资源个数	二维码
电子教案	Word 文档	8	
教学课件	PPT 文档	8	
教学视频	MP4 视频	98	
课后习题	Word 文档及源码	8	

在本书的成稿与出版过程中，出版社的编辑同志敬业负责，还有很多同行及专家提出了许多的宝贵意见，在此，对所有为本书提供过帮助的同事和朋友们表示衷心的感谢！由于编者水平有限，书中难免有不妥之处，敬请各位读者与专家批评指正。

编　者

2023 年 3 月

目　　录

项目一　Java Web 开发环境准备 ……… 1
任务 1　开发环境搭建 ………………… 2
任务演示 …………………………… 2
知识准备 …………………………… 2
1. Java Web 的基本概念 …………… 2
2. JDK 安装与配置 ………………… 2
3. Tomcat 安装与配置 ……………… 6
4. Eclipse 下载与 Tomcat 配置 …… 10
任务实施 …………………………… 14
任务拓展 …………………………… 17
1. Tomcat 启动 ……………………… 17
2. Tomcat 关闭 ……………………… 18
3. UTF-8 编码设置 ………………… 18

任务 2　输出"欢迎进入智慧金融信贷管理系统" ………………… 19
任务演示 …………………………… 19
知识准备 …………………………… 20
1. JSP 页面简介 …………………… 20
2. JSP 的运行原理 ………………… 21
任务实施 …………………………… 21
任务拓展 …………………………… 22
1. C/S 架构 ………………………… 22
2. B/S 架构 ………………………… 22

项目实训 ……………………………… 23
实训一　在计算机上搭建 Java Web 开发环境 ……………………… 23
实训二　输出词《沁园春·雪》 …… 23

课后练习 ……………………………… 23

项目二　JSP 技术 ……………………… 25
任务 1　化妆品网站框架设计 ………… 26
任务演示 …………………………… 26
知识准备 …………………………… 27
1. `<font></font>` 字体标签 ………… 27
2. `<hn></hn>` 标题标签 …………… 28
3. `<div>` 分区标签 ………………… 28
4. `<p>` 段落标签 …………………… 29
5. `<br>` 换行标签 …………………… 29
6. `<hr>` 水平分隔线标签 …………… 29
7. `<B></B>`、`<I></I>`、`<U></U>` 标签 … 29
8. `<img>` 标签 ……………………… 30
9. 超链接标签 ……………………… 30
10. `<table></table>` 表格标签 ……… 31
任务实施 …………………………… 32
任务拓展 …………………………… 33
1. `<head></head>` 标签与 `<title></title>` 标签 …………………………… 33
2. `<meta>` 标签 …………………… 34

任务 2　温馨提示语定时显示程序设计 …………………………… 35
任务演示 …………………………… 35
知识准备 …………………………… 36
1. 什么是 JSP ……………………… 36
2. JSP 页面的基本结构 …………… 36
3. JSP 程序的脚本元素 …………… 37

4. JSP 的注释 ································ 40
　任务实施 ·· 41
　任务拓展 ·· 42
任务 3　美景欣赏网站设计 ················ 42
　任务演示 ·· 42
　知识准备 ·· 43
　　1. JSP 的指令标识 ···························· 43
　　2. JSP 动作元素 ······························· 46
　任务实施 ·· 50
　任务拓展 ·· 54
项目实训 ··· 55
　实训一　设计中秋节网站框架 ········ 55
　实训二　设计信贷数据分析可视化
　　　　　平台用户注册界面 ············ 55
课后练习 ··· 56

项目三　JSP 内置对象 ················ 58
任务 1　应用 request 对象设计网上
　　　　考试系统 ································ 59
　任务演示 ·· 59
　知识准备 ·· 59
　　1. out 内置对象 ······························· 60
　　2. <form></form>表单标记 ··········· 61
　　3. request 对象 ······························· 62
　　4. request 对象的常用方法 ············ 64
　任务实施 ·· 66
　任务拓展 ·· 68
　　1. 表单提交的 method 方法 ·········· 68
　　2. 解决中文乱码问题 ···················· 69
任务 2　应用 response 对象设计化妆品
　　　　网站登录界面 ························ 69
　任务演示 ·· 69
　知识准备 ·· 70
　　1. response 对象重定向 ················· 70
　　2. response 对象刷新页面 ············· 72
　任务实施 ·· 72

　任务拓展 ·· 76
任务 3　应用 session 对象设计火锅
　　　　点餐系统 ································ 77
　任务演示 ·· 77
　知识准备 ·· 77
　　1. session 对象 ······························· 77
　　2. session 对象的 id ······················· 78
　　3. session 对象的常用方法 ············ 78
　任务实施 ·· 80
　任务拓展 ·· 82
任务 4　应用 application 对象设计
　　　　留言板 ···································· 83
　任务演示 ·· 83
　知识准备 ·· 83
　　1. application 对象 ························· 83
　　2. Vector 类 ···································· 84
　　3. <textarea></textarea>标记 ········· 85
　任务实施 ·· 85
　任务拓展 ·· 86
任务 5　应用 Cookie 对象制作站点
　　　　计数器 ···································· 89
　任务演示 ·· 89
　知识准备 ·· 89
　　1. Cookie 对象的创建 ···················· 89
　　2. Cookie 对象的读取 ···················· 89
　　3. Cookie 的常用方法 ···················· 90
　任务实施 ·· 91
　任务拓展 ·· 92
项目实训 ··· 92
　实训一　根据家庭生活采购账单
　　　　　计算消费总额 ···················· 92
　实训二　设计调查问卷主界面 ········ 93
　实训三　设计用户注册程序 ············ 93
课后练习 ··· 93

项目四　JDBC 与数据库访问 ············ 96
任务 1　Java 程序连接数据库 ············· 97

任务演示	97
知识准备	97
1. JDBC 简介	97
2. JDBC 的类和接口	98
任务实施	103
任务拓展	111

任务 2　数据库查询和模糊查询 113

任务演示	113
知识准备	114
1. 使用 PreparedStatement 接口实现查询	114
2. 使用 PreparedStatement 接口实现模糊查询	114
任务实现	115
任务拓展	117
1. 静态代码块	117
2. 工具类 DBUtils	118

任务 3　数据库更新 119

任务演示	119
知识准备	120
任务实施	120
任务拓展	123

任务 4　应用数据库连接池驱动 MySQL 数据库 125

任务演示	125
知识准备	126
1. 使用数据库连接池的必要性	126
2. 数据库连接池技术	126
项目实施	128
任务拓展	129

项目实训 131

实训　创建数据库并进行相关的数据库操作	131
课后练习	131

项目五　JavaBean 开发模型 133

任务 1　应用 JavaBean 计算梯形的面积 134

任务演示	134
知识准备	134
1. JavaBean 的简介	134
2. JavaBean 的分类	134
3. JavaBean 的规范	134
4. JavaBean 的应用	135
任务实施	137
任务拓展	139
1. FilenameFilter 介绍	139
2. File 类	139

任务 2　应用 JavaBean 实现化妆品网站注册功能 141

任务演示	141
知识准备	142
任务实施	143
任务拓展	145

任务 3　应用 JavaBean 实现水果购物车系统 147

任务演示	147
知识准备	148
任务实施	148
任务拓展	153

项目实训 157

实训一　应用 JavaBean 技术设计留言板	157
实训二　应用 JavaBean 技术实现四则运算	157
实训三　为化妆品网站设计查看功能	158
课后练习	158

项目六　Servlet 技术 160

任务 1　创建并配置 Servlet 程序 161

任务演示 …………………………… 161	1. 认识 Listener …………………… 211
知识准备 …………………………… 162	2. 创建 Listener …………………… 212
1. Servlet 的简介 ………………… 162	项目实训 …………………………… 216
2. Servlet 的特点 ………………… 162	实训一 统计站点的访问次数 …… 216
3. Servlet 的生命周期 …………… 162	实训二 设计用户注册界面 ……… 216
4. Servlet 的接口 ………………… 163	实训三 设计过滤器 ……………… 217
5. HttpServlet 的类 ……………… 164	课后练习 …………………………… 217
6. Servlet 的配置 ………………… 164	**项目七 EL 和 JSTL 技术** ………… 219
任务实施 …………………………… 165	任务 1 应用 JSTL 实现用户个人
任务拓展 …………………………… 171	信息获取 …………………… 220
1. ServletConfig 对象 …………… 171	任务演示 …………………………… 220
2. ServletContext 对象 ………… 172	知识准备 …………………………… 221

任务 2 应用 HttpServletRequest 对象
和 HttpServletResponse 对象
实现用户验证登录 ………… 177
任务演示 …………………………… 177
知识准备 …………………………… 177
　1. HttpServletRequest 对象……… 177
　2. HttpServletResponse 对象 …… 185
　3. 请求重定向 …………………… 188
任务实施 …………………………… 191
任务拓展 …………………………… 191
　1. 解决中文乱码问题 …………… 191
　2. 使用 HttpServletRequest 对象传递
　　 数据 …………………………… 194
　3. 请求转发 ……………………… 194
任务 3 应用 Commons-FileUpload 和
Filter 实现文件上传与下载 … 195
任务演示 …………………………… 195
知识准备 …………………………… 196
　1. 文件上传流程 ………………… 196
　2. Commons-FileUpload 组件 …… 197
　3. 文件下载流程 ………………… 198
　4. Filter ………………………… 199
任务实施 …………………………… 200
任务拓展 ……………………………211

　1. EL 表达式 ……………………… 221
　2. EL 表达式语法 ………………… 221
　3. EL 表达式标识符 ……………… 223
　4. EL 关键字 ……………………… 223
　5. EL 变量与常量 ………………… 223
　6. EL 访问数据 …………………… 224
　7. EL 运算符 ……………………… 224
　8. EL 运算符的优先级 …………… 226
　9. 使用 EL 表达式从作用域中获取
　　 数据 …………………………… 226
　10. EL 的隐式对象 ………………… 227
任务实施 …………………………… 232
任务拓展 …………………………… 234
　1. 应用 EL 获取 Cookie 对象的信息 …… 234
　2. 应用 EL 获取 initParam 对象的信息 … 234
任务 2 应用 JSTL 实现商品展示 …… 235
任务演示 …………………………… 235
知识准备 …………………………… 235
　1. JSTL 的概念 …………………… 235
　2. 下载和安装 JSTL ……………… 237
　3. JSTL 的核心标签库 …………… 239
任务实施 …………………………… 247
任务拓展 …………………………… 249

项目实训 252
 实训一 应用 EL 表达式显示"好好学习，天天向上！" 252
 实训二 应用 EL 表达式设置页面的背景色 252
 实训三 遍历集合中的元素 252
 实训四 应用 EL 表达式得到一个计算器 252
课后练习 253

项目八 智慧金融信贷管理系统 255

任务 1 智慧金融信贷管理系统搭建 256
 任务演示 256
 知识准备 258
  1. 系统整体架构 258
  2. 智慧金融信贷管理系统数据库设计 259
 任务实施 260

任务 2 智慧金融信贷管理系统注册功能实现 265
 任务演示 265
 知识准备 265
  1. 功能描述 265
  2. 注册功能类 265
  3. 注册界面的功能时序 266
 任务实施 266

任务 3 智慧金融信贷管理系统登录功能实现 271
 任务演示 271
 知识准备 271
  1. 功能描述 271
  2. 登录功能类 272
  3. 登录界面的功能时序 272
 任务实施 272

任务 4 智慧金融信贷管理系统贷款申请功能实现 280
 任务演示 280
 知识准备 280
  1. 功能描述 280
  2. 贷款申请功能类 281
  3. 贷款申请的功能时序 281
 任务实施 282

任务 5 管理员登录功能实现 287
 任务演示 287
 知识准备 287
  1. 功能描述 287
  2. 管理员登录功能类 287
  3. 管理员登录的功能时序 288
 任务实施 288

任务 6 贷款用户信息查询功能实现 293
 任务演示 293
 知识准备 294
  1. 功能描述 294
  2. 贷款用户功能类 294
  3. 贷款用户的功能时序 294
 任务实施 296

项目一　Java Web 开发环境准备

项目要求

本项目内容是 Java Web 开发环境准备,要求完成 Java Web 开发环境搭建,并能编写、编译、运行 Java Web 程序。

项目分析

要完成项目任务,至少需要具备两个基本条件:一是需要在计算机上安装和配置 JDK,二是要在计算机上安装 JSP 引擎,如 Tomcat 服务器等。该项目分为两个任务,分别是开发环境搭建和输出"欢迎进入智慧金融信贷管理系统"。

项目目标

【知识目标】熟悉 Java Web 的基本概念,掌握 JDK 和 Tomcat 的安装与配置,以及 JSP 运行环境的配置。

【能力目标】能编写、编译简单的 JSP 程序。

【素质目标】提高学生发现问题、分析问题、解决问题的能力。

知识导图

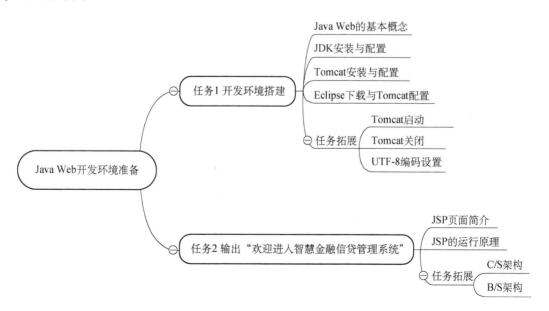

任务 1　开发环境搭建

任务演示

本任务要完成 JDK 的安装与配置、Tomcat 安装与配置、Eclispe 下载与 Tomcat 配置，为输出"欢迎进入智慧金融信贷管理系统"做好准备。如图 1-1 所示为开发环境搭建成功后的 JSP 运行界面。

图 1-1　JSP 运行界面

知识准备

1. Java Web 的基本概念

Java Web 是用 Java 技术来解决 Web 互联网领域的问题的技术栈，主要包括 Servlet、JSP、JavaBean、JDBC 等技术。目前，Java Web 动态网站已经被广泛应用于电子商务、电子政务、网络资源管理及大数据等领域。

2. JDK 安装与配置

Java Web 应用程序开发离不开 JDK（Java Development Kit）和 JRE（Java Runtime Enviroment），JDK 是 Java 语言的编译环境，JRE 是 Java 的运行环境，必须安装 JDK 和 JRE，并设置相应环境变量，才可以编译和执行 Java 程序。下面介绍 JDK 安装与配置。

教学视频

（1）JDK 下载

可以从 Oracle 官网免费下载 JDK，只需先在浏览器地址栏输入 Oracle 官网网址，然后在打开的官网首页单击"Developers"按钮，如图 1-2 所示，即可进入下载界面。

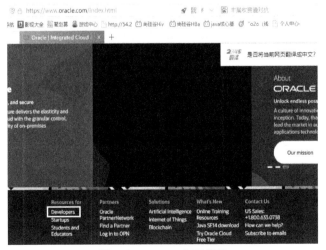

图 1-2　Oracle 官网首页

在下载界面单击"Downloads"下拉列表,选择"Java SE"选项,如图 1-3 所示。

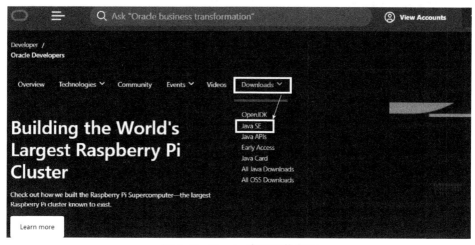

图 1-3　Oracle 官网下载界面

在进入的 JDK 下载界面中单击"Java SE 8"下方的"Download"按钮进行下载,如图 1-4 所示。

图 1-4　JDK 下载界面 1

JDK 下载界面显示了可供下载的、支持各种操作系统的 JDK,可根据操作系统的位数下载合适的 JDK(32 位 Windows 操作系统选择 Windows x86 版本,64 位 Windows 操作系统选择 Windows x64 版本),如图 1-5 所示。

(2)安装 JDK

JDK 安装步骤如下。

第一步:双击已下载的 JDK 文件,在弹出的对话框中选择接受许可证协议,进入安装程序界面,如图 1-6 所示。

图 1-5　JDK 下载界面 2

图 1-6　安装程序界面

第二步：单击"下一步"按钮进入定制安装界面，默认安装路径是"C:\Program Files\Java\jdk1.8.0_77\"，这里选择默认安装路径，当然也可单击"更改"按钮更改安装路径，定制安装界面如图 1-7 所示。

图 1-7　定制安装界面

【脚下留心】
　　注意，安装路径中不要出现中文。

第三步：单击"下一步"按钮，进入 JDK 安装完成界面，如图 1-8 所示。

图 1-8　JDK 安装完成界面

（3）JDK 环境配置

安装好 JDK 后，还需要设置一个 JAVA_HOME 环境变量，使它指向 JDK 的安装路径，基本步骤如下。

第一步：在桌面上右键单击"我的电脑"桌面图标，选择"属性"选项，在打开的对话框中选择"高级系统设置"选项卡，单击"环境变量"按钮打开"环境变量"对话框，如图 1-9 所示。

图 1-9　"环境变量"对话框

第二步：在"编辑系统变量"对话框中增加系统变量，变量名为"JAVA_HOME"，变量值设置为 JDK 的安装路径"C:\Program Files\Java\jdk1.8.0_77"，如图 1-10 所示。

图 1-10 "编辑系统变量"对话框

第三步：退出"编辑系统变量"对话框，查看系统变量中是否有 Path 变量，若没有则在"编辑环境变量"对话框中添加"JAVA_HOME\bin"和"JAVA_HOME%\jre\bin"，同时添加安装路径"C:\Program Files\Java\jdk1.8.0_77\bin"，如图 1-11 所示。

图 1-11 设置系统变量 Path

系统变量设置完成后，还需验证 JDK 是否安装和设置成功。从开始菜单打开运行窗口，输入"cmd"打开命令提示符窗口，在打开的命令提示符窗口中输入"javac"。如果 JDK 安装和设置成功，则会出现如图 1-12 所示的信息。

图 1-12 验证 JDK 是否安装和设置成功

3. Tomcat 安装与配置

（1）Tomcat 下载

Tomcat 是 Apache 软件基金会的 Jakarta 项目中的一个重要子项目，具有免费和跨平台的

教学视频

特点，并且运行稳定、可靠性强、效率高，是使用最广泛的 Servlet/JSP 服务器。用户可以通过官方网站下载 Tomcat，下载界面如图 1-13 所示。

图 1-13　Tomcat 的下载界面

在下载界面左侧的"Download"按钮下方选择要下载的版本。这里以 Tomcat8.0 为例，在确定下载的版本后，选择 Tomcat8 的子版本 Tomcat8.0，将鼠标光标往下移，找到"Binary Distributions"下的"core"，并单击"zip(pgp，sha512)"按钮下载所需要的 Tomcat，如图 1-14。

图 1-14　Tomcat8.0 的下载页面

（2）Tomcat 安装与配置

成功下载 Tomcat 的可执行文件后，即可安装 Tomcat。

第一步：先双击 apache-tomcat-8.0.32.exe 可执行文件，安装界面如图 1-15 所示，然后单击"Next"按钮继续安装。

第二步：进入选择安装组件界面，如图 1-16 所示，在选中"Full"选项后，单击"Next"按钮继续安装。

第三步：打开配置界面，指定 Tomcat 的 HTTP 端口号，默认是 8080，如图 1-17 所示，单击"Next"按钮继续安装。

图 1-15　安装界面

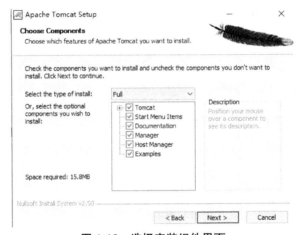

图 1-16　选择安装组件界面

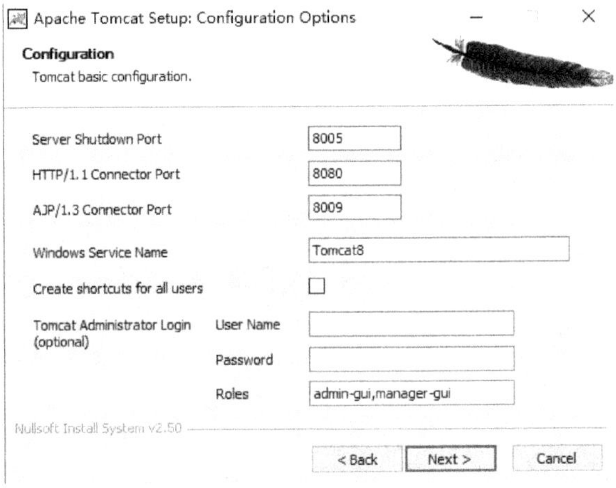

图 1-17　配置界面

第四步：正在安装的界面如图 1-18 所示。

Tomcat 安装完成后，还不能直接使用，还需要进行 Tomcat 环境配置。在此需要记住安装路径，因为 Tomcat 要配置和 JDK 一样的环境变量。同样在"编辑系统变量"对话框中

图 1-18　正在安装的界面

增加系统变量，设置变量名为"CATALINA_HOME"，设置变量值为 Tomcat 的解压路径，如图 1-19 所示。

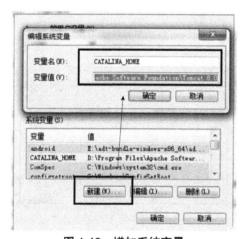

图 1-19　增加系统变量

（3）Tomcat 的目录结构

Tomcat 的安装目录（即安装路径）中包含了一系列的子目录，这些子目录分别用于存放不同功能的文件，Tomcat 的目录结构如图 1-20 所示。

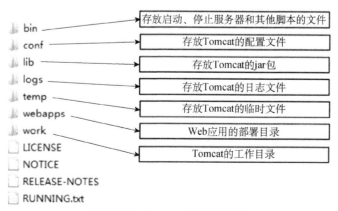

图 1-20　Tomcat 的目录结构

（4）Tomcat 启动与停止

在开始菜单中选中"Configure Tomcat"选项，如图 1-21 所示，随即进入"Tomcat8 Properties"对话框，并单击"Start"按钮启动 Tomcat，如图 1-22 所示。

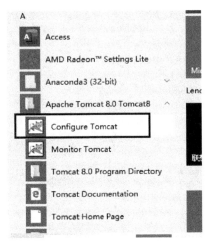

图 1-21　选中"Configure Tomcat"选项

图 1-22　"Tomcat8 Properties"对话框

在成功安装和启动 Tomcat 后，在浏览器的地址栏中输入"localhost:8080"，如果出现如图 1-23 所示的 Tomcat 主界面，则表示 Tomcat 的安装和配置正常。

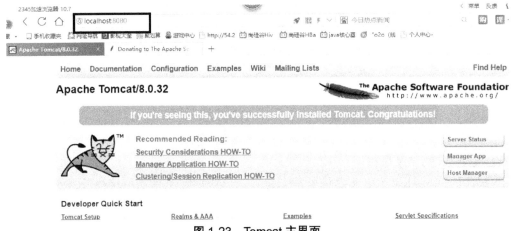

图 1-23　Tomcat 主界面

4. Eclipse 下载与 Tomcat 配置

（1）Eclipse 下载

Eclipse 是一个开放的、可扩展的集成开发环境。它不仅可以用于 Java 桌面程序开发，而且在安装了开发插件后，还可以构建 Web 项目和移动项目的开发环境。Eclipse 是一个开放源代码的项目，可以免费从官方网站下载，下载后的 Eclipse 可直接解压使用，Eclipse 下载界面如图 1-24 所示。

教学视频

（2）在 Eclipse 中配置 JRE

在 Eclipse 工作界面中依次选择"Window"→"Preferences"选项，打开如图 1-25 所示的"Preferences"对话框，展开"Preferences"对话框左侧的"Java"节点，选择该节点下的

"Installed JREs"子节点。如果"Preferences"对话框右侧的名称、位置与安装的 JRE 是一致的，则单击"OK"按钮。如果不一致，则需要修改。

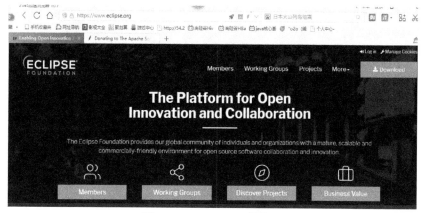

图 1-24　Eclipse 下载界面

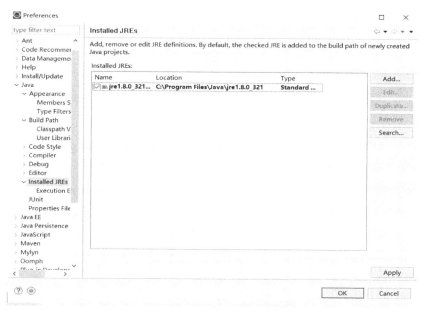

图 1-25　"Preferences"对话框

（3）在 Eclipse 中配置 Tomcat

第一步：启动 Eclipse，依次选择"Window"→"Preferences"选项，打开"Preferences"对话框，展开"Preferences"对话框左侧的"Server"节点下的"Runtime Environments"子节点，如图 1-26 所示。在"Preferences"对话框的右侧选择 Tomcat 的安装目录，单击"OK"按钮。

第二步：单击图 1-26 中的"Add"按钮打开"New Server Runtime Environment"对话框，该对话框展示了可在 Eclipse 中配置的各种服务器及其版本。由于使用 apache-tomcat-8.0.32.exe 进行安装，因此在此选择"Apache Tomcat v8.0"，如图 1-27 所示。

第三步：在"New Server Runtime Environment"对话框中单击"Next"按钮，在打开的"Tomcat Server"对话框中单击"Browse"按钮选择 Tomcat 的安装目录，如图 1-28 所示。

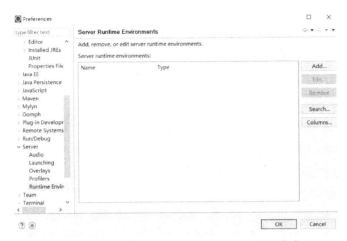

图 1-26 展开 "Runtime Environments" 子节点

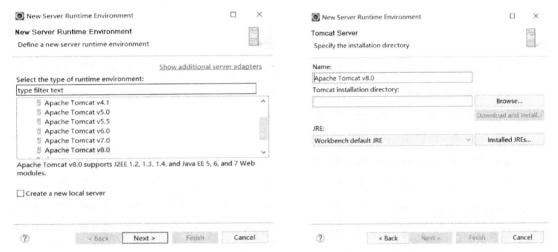

图 1-27 "New Server Runtime Environment" 对话框　　　图 1-28 选择 Tomcat 的安装目录

第四步：在选择安装目录以后，单击"OK"按钮，如图 1-29 所示。

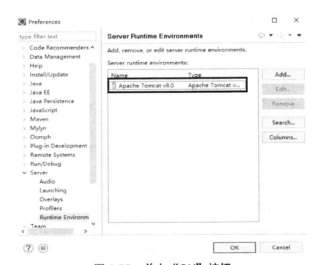

图 1-29 单击 "OK" 按钮

第五步：选择"Servers"选项，创建新的 Server 窗口，如图 1-30 所示。

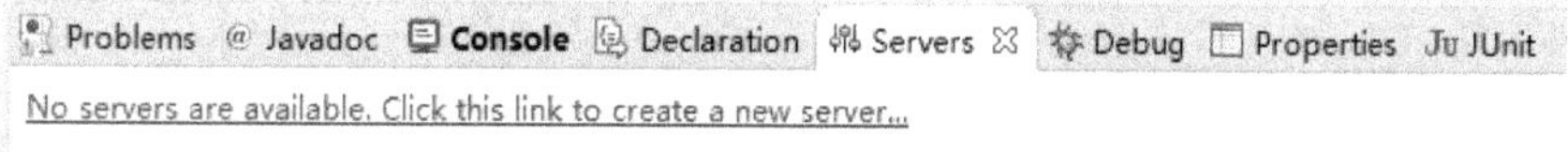

图 1-30　创建新的 Server 窗口

【多学一招】
　　如果没有"Servers"选项，则可以依次选择"Window"→"Show View"选项打开"Servers"选项。

第六步：单击图 1-30 上的超链接，会弹出一个如图 1-31 所示的"New Server"对话框，选中"Tomcat v8.0 Server"选项，单击"Finish"按钮。

图 1-31　"New Server"对话框

到此就完成了 Tomcat 的配置，根据图 1-32 上的标记依次单击相应选项就可以启动 Tomcat。

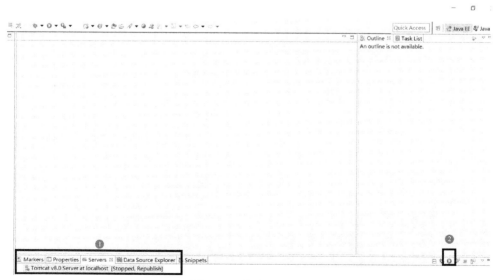

图 1-32　启动 Tomcat

任务实施

在搭建好的 Java Web 开发环境中创建一个 JSP 文件，并输出"今天太高兴了，Java Web 开发环境终于搭建成功"。

第一步：在 Eclipse 中依次单击"File"→"New"→"Dynamic Web Project"选项，打开如图 1-33 所示的"New Dynamic Web Project"对话框，在此对话框的"Project name"后方输入：Test。

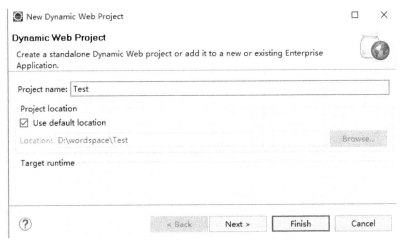

图 1-33 "New Dynamic Web Project"对话框

单击"Next"按钮，直到出现"Web Module"对话框，在该对话框中选择如图 1-34 所示的复选框。

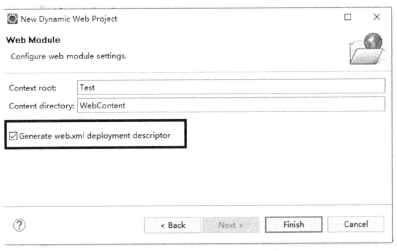

图 1-34 "Web Module"对话框

第二步：新建一个 index.jsp 文件，如图 1-35 所示。在"File name"后的文本框内输入：index.jsp，单击"Finish"按钮完成文件创建。

第三步：在<body></body>标签内输入：今天太高兴了，Java Web 开发环境终于搭建成功。参考代码如下。

```
<%@ page language="java" contentType="text/html;charset=UTF-8"
    pageEncoding="UTF-8"%>
<!DOCTYPE html PUBLIC "-//W3C//DTD HTML 4.01 Transitional//EN" "http://www.w3.org/TR/html4/loose.dtd">
<html>
<head>
<meta http-equiv="Content-Type" content="text/html;charset=UTF-8">
<title>Insert title here</title>
</head>
<body bgcolor="aaffee">
今天太高兴了,Java Web 开发环境终于搭建成功
</body>
</html>
```

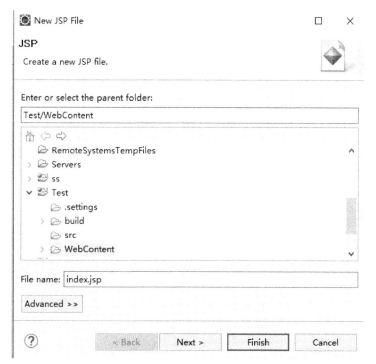

图 1-35　新建一个 index.jsp 文件

代码解析：

① language="java"：表示使用的语言，这里只能是 Java 语言；

② pageEncoding="utf-8"：表示字符编码格式。

补充说明：JSP 页面将以文本文件的形式保存，扩展名是.jsp。在保存 JSP 页面时，文件名必须符合标识符规定，即文件名可以由字母、下画线、美元符号和数字组成，第一个字符不能是数字，并严格区分字母大小写，比如"Test.jsp"和"test.jsp"是不相同的。

第四步：运行 Java Web，右键单击"Test"项目，将鼠标光标移动到"Run As"选项上，然后选中"Run on Server"选项，如图 1-36 所示。

单击图 1-37 中的"Finish"按钮，选择要运行的"Test"项目，如图 1-38 所示。

运行结果如图 1-39 所示。

16 / Java Web 程序设计项目实战（微课版）

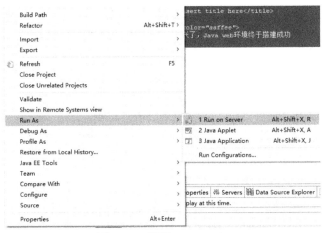

图 1-36 运行 Java Web

图 1-37 单击"Finish"按钮

图 1-38 选择要运行的"Test"项目

图 1-39 运行结果

任务拓展

1. Tomcat 启动

启动 Tomcat 的方法有两种，下面分别对其进行介绍。

（1）方法一

① 通过位于目录"D:\Program Files\Apache Software Foundation\Tomcat 8.0\bin"下的 startup.bat 启动 Tomcat。

② 打开浏览器输入：http://localhost:8080，或者输入：http://127.0.0.1:8080，即可访问 Tomcat，如果显示如图 1-40 所示的信息，则表示 Tomcat 启动成功。

图 1-40 Tomcat 启动成功

（2）方法二

使用 catalina.bat 启动 Tomcat，如图 1-41 所示。

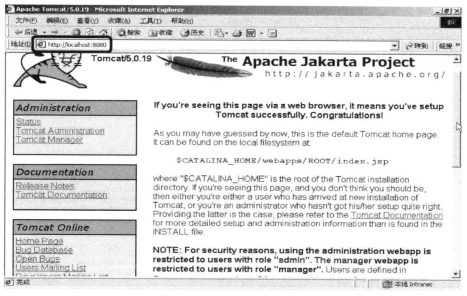

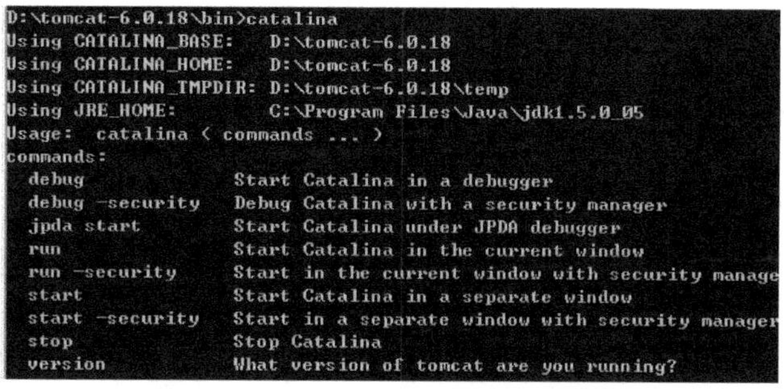

图 1-41 使用 catalina.bat 启动 Tomcat

【多学一招】

8080 是 Tomcat 的默认端口号，如果 Tomcat 使用的端口号已经被占用，则 Tomcat 将无法启动。我们可以通过修改 Tomcat 安装目录中的 conf 文件夹下的配置文件 server.xml 来更改端口号。用记事本打开 server.xml 文件，找到如图 1-42 所示的方框中的代码，将 “port="8080"” 中的端口号更改为新的端口号（比如将 8080 更改为 9090 等），并重新启动 Tomcat。

```
<!--
<Connector executor="tomcatThreadPool"
       port="8080" protocol="HTTP/1.1"
       connectionTimeout="20000"
       redirectPort="8443" />
-->
<!-- Define a SSL/TLS HTTP/1.1 Connector on port 8443
   This connector uses the NIO implementation that requires the JSSE
   style configuration. When using the APR/native implementation, the
   OpenSSL style configuration is required as described in the APR/native
   documentation -->
```

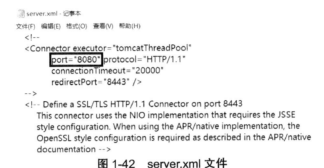

图 1-42　server.xml 文件

2. Tomcat 关闭

进入 startup.bat 的同级目录，双击 shutdown.bat 文件关闭 Tomcat，如图 1-43 所示。

名称	修改日期	类型	大小
bootstrap.jar	2016-02-03 3:38	Executable Jar File	28 KB
catalina.bat	2016-02-03 3:38	Windows 批处理文件	14 KB
configtest.bat	2016-02-03 3:38	Windows 批处理文件	2 KB
digest.bat	2016-02-03 3:38	Windows 批处理文件	3 KB
service.bat	2016-02-03 3:38	Windows 批处理文件	7 KB
setclasspath.bat	2016-02-03 3:38	Windows 批处理文件	4 KB
shutdown.bat	2016-02-03 3:38	Windows 批处理文件	2 KB
startup.bat	2016-02-03 3:38	Windows 批处理文件	2 KB
tcnative-1.dll	2016-02-03 3:38	应用程序扩展	2,076 KB
Tomcat8.exe	2016-02-03 3:38	应用程序	108 KB
Tomcat8w.exe	2016-02-03 3:38	应用程序	108 KB
tomcat-juli.jar	2016-02-03 3:38	Executable Jar File	40 KB
tool-wrapper.bat	2016-02-03 3:38	Windows 批处理文件	4 KB
version.bat	2016-02-03 3:38	Windows 批处理文件	2 KB

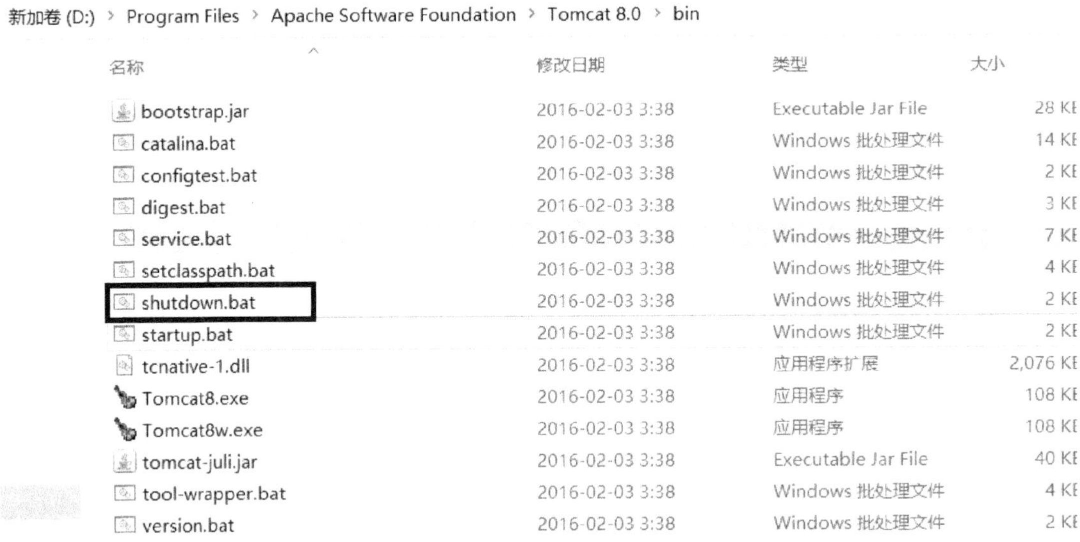

图 1-43　关闭 Tomcat

3. UTF-8 编码设置

JSP 默认的编码格式是 ISO-8859-1，该编码不能识别中文，如图 1-44 所示。在此需要将编码格式修改为能识别中文的 UTF-8。

依次选择 “Window” → “Preferences” 选项，在搜索框中输入：jsp，选择 “JSP Files” 节点，在 “Encoding” 后面选择 “ISO 10646/Unicode（UTF-8）” 选项，即编码格式为 UTF-8，如图 1-45 所示。

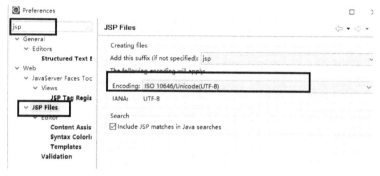

图 1-44　ISO-8859-1 不能识别中文

图 1-45　UTF-8 编码设置

新建 JSP 文件，发现字符编码自动修改为 UTF-8，如图 1-46 所示。

图 1-46　字符编码自动修改后的结果

任务 2　输出"欢迎进入智慧金融信贷管理系统"

任务演示

本次任务在 Java Web 开发环境搭建成功的基础上，编写能输出"欢迎进入智慧金融信贷管理系统"的网站，并编译和发布此网站，如图 1-47 所示。

图 1-47　输出"欢迎进入智慧金融信贷管理系统"

知识准备

1. JSP 页面简介

JSP 页面中可以有普通的 HTML 标记和 JSP 规定的 JSP 标记,以及通过标记符号("<%"和"%>")加入的 Java 程序。JSP 页面以文本文件的形式保存,扩展名是.jsp。在保存 JSP 页面时,文件名必须符合标识符规定,即文件名可以由字母、下画线、美元符号和数字组成,并且第一个字符不能是数字。

> 【脚下留心】
> JSP 是基于 Java 语言的技术,其文件名区分字母大小写,比如 hello.jsp 和 Hello.jsp 是不同的 JSP 文件。

【例 1-1】计算 1~100 的所有整数的累加和,参考代码如下,运行结果如图 1-48 所示。

```jsp
1  <%@ page language="java" contentType="text/html;charset=UTF-8"
2  pageEncoding="UTF-8"%>
3  <!DOCTYPE html PUBLIC "-//W3C//DTD HTML 4.01 Transitional//EN" "http://www.w3.org/TR/html4/loose.dtd">
4  <html>
5  <head>
6  <meta http-equiv="Content-Type" content="text/html;charset=UTF-8">
7  <title>Insert title here</title>
8  </head>
9  <body bgcolor="pink">
10 <h2>计算1~100之和</h2>
11 <% int i,sum=0;
12 for(i=1;i<=100;i++)
13 {
14 sum=sum+i;
15 }
16 %>
17 <p>1 到 100 之和是:</p>
18 <%=sum %>
19 </body>
20 </html>
```

图 1-48 计算 1~100 的所有整数的累加和的运行结果

代码说明:

第 1~2 行是 JSP 的声明;第 4~19 行是 HTML 语言,其中第 11~16 行是 Java 脚本,第 18 行是 JSP 表达式。

2. JSP 的运行原理

当 Tomcat 上的 JSP 文件被第一次请求执行时，Tomcat 上的 JSP 引擎首先将 JSP 文件转译成 Java 文件，并编译 Java 文件生成字节码文件，然后执行字节码文件来响应客户的请求。

① 把 JSP 文件中的 HTML 标记（即页面的静态部分）交给浏览器显示。
② 负责处理 JSP 标记，并将处理结果发送到浏览器。
③ 执行 "<%" 和 "%>" 之间的 Java 程序（即页面的动态部分），并把执行结果交给浏览器显示。
④ 当多个用户请求一个 JSP 文件时，Tomcat 会为每个用户启动一个线程，该线程负责执行常驻内存的字节码文件来响应相应用户的请求。

> 【脚下留心】
> 如果对 JSP 文件进行了修改、保存，那么 Tomcat 会执行新的字节码文件。

所有 JSP 文件在被执行时都会被服务器端的 JSP 引擎转译为 Java 文件，又由 JSP 引擎调用 Java 编译器将 Java 文件编译为 class 文件，并由 Java 虚拟机（JVM）解释和执行，JSP 的运行原理如图 1-49 所示。

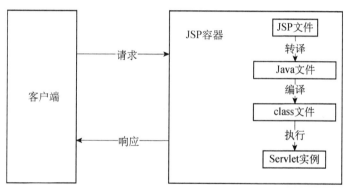

图 1-49　JSP 的运行原理

任务实施

编写能输出"欢迎进入智慧金融信贷管理系统"的网站，并编译和发布此网站，参考代码如下，运行结果如图 1-47 所示。

```jsp
<%@ page language="java" contentType="text/html;charset=UTF-8"
    pageEncoding="UTF-8"%>
<!DOCTYPE html PUBLIC "-//W3C//DTD HTML 4.01 Transitional//EN" "http://www.w3.org/TR/html4/loose.dtd">
<html>
<head>
<meta http-equiv="Content-Type" content="text/html;charset=UTF-8">
<title>Insert title here</title>
</head>
<body bgcolor="aaffee">
```

```
    <b>欢迎进入智慧金融信贷管理系统</b>
</body>
</html>
```

任务拓展

在进行软件开发时，通常会在两种软件架构中进行选择，即 C/S 架构和 B/S 架构。C/S 架构是客户端（Client）/服务器（Server）的交互；B/S 架构是浏览器（Browser）/服务器（Server）的交互。

1. C/S 架构

C/S 架构是一种出现得比较早的软件架构。在 C/S 架构中，多个客户端程序可以同时访问一个 Web 服务器或数据库服务器，如图 1-50 所示。"C"表示客户端，用户通过客户端使用软件。"S"表示服务器，服务器用来处理软件的业务逻辑。常见的采用 C/S 架构的软件有 QQ、微信等。

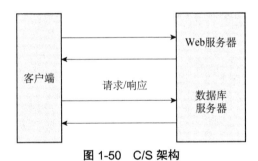

图 1-50　C/S 架构

C/S 架构的特点如下。
① 必须在软件使用前安装。
② 软件更新时，服务器和客户端需同时更新。
③ 使用 C/S 架构的软件不支持跨平台。
④ 使用 C/S 架构的软件客户端和服务器进行通信是相对安全的。

2. B/S 架构

随着 Internet 技术的兴起，诞生了一种新的软件架构——B/S 架构，它是 Web 兴起后出现的一种网络架构模式，是 Web 服务器与数据库交互的过程，如图 1-51 所示。"B"表示浏览器，"S"表示服务器，常见的使用浏览器访问网页的软件有京东、淘宝、知乎、新浪微博等。

图 1-51　B/S 架构

B/S 架构的特点如下。
① 不需要安装软件，可直接使用浏览器访问指定的网址。
② 软件更新时，客户端不需要更新。
③ 软件可以跨平台，只要系统中有浏览器就可以使用。
④ 使用 B/S 架构的软件客户端与服务器进行通信不是很安全。

项 目 实 训

实训一　在计算机上搭建 Java Web 开发环境

请根据项目一所学内容，在计算机上搭建 Java Web 开发环境。

实训二　输出词《沁园春·雪》

要求：应用 Eclipse 开发工具，输出词《沁园春·雪》，输出效果如图 1-52 所示。

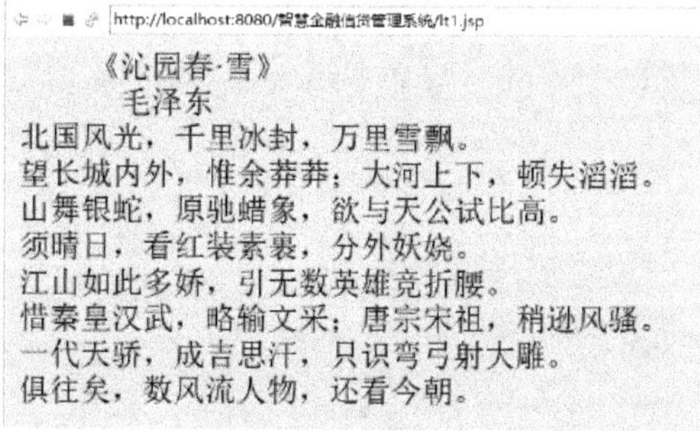

图 1-52　输出效果

课 后 练 习

一、填空题

1. Tomcat 的默认端口号是_____。
2. JSP 文件的扩展名是_____。
3. JSP 页面默认使用的程序语言是_____。
4. 启动 Tomcat 时，可以在浏览器输入的地址是_____。
5. JSP 默认的编码方式是_____。
6. 在进行软件开发时，通常会在两种软件架构中进行选择，即_____架构和_____架构。

二、选择题

1. 在关闭 Tomcat 时，可双击（　　）文件关闭 Tomcat。

A. shutdown.bat　　　B. configtest.bat　　　C. digest.bat　　　D. service.bat

2. 下列不符合 JSP 标识符规定的是（　　）。

A. System1　　　B. studnt_info　　　C. Main　　　D. 8ab_

3. Java 是一种非常高效的编程语言，下列不属于其特性的是（　　）。

A. 简单、跨平台、分布式　　　　　　B. 健壮性和安全性
C. 基于对象、解释运行　　　　　　　D. 多线程、网络功能强大

4. 在 Java 源程序代码中，可使用（　　）语句把当前文件放入指向的包中。

A. import　　　B. public class　　　C. package　　　D. interface

5. 在静态 Web 中，下列说法错误的是（　　）。

A. 在静态 Web 中可以插入 GIF 动画图片
B. 在静态 Web 中可以插入 JavaScript 代码
C. 在静态 Web 中可以插入 Java 代码
D. 在静态 Web 中可以插入 Flash 动画

三、简答题

1. Boy.jsp 和 boy.jsp 是否是相同的 JSP 文件名？
2. 请在 server.xml 文件中将端口号修改为 9999。
3. 请启动 Tomcat，如果已经启动，则必须先关闭 Tomcat，再重新启动。

项目二 JSP 技术

项目要求

本项目要完成 JSP 技术应用，主要完成静态网站的框架设计，并应用 JSP 基础语法编写简单的 JSP 应用程序。

项目分析

要完成项目任务，至少需要具备两个基本条件：一是熟悉 HTML 常用标记，二是掌握 JSP 基础语法。现将该项目分为三个任务，分别是化妆品网站框架设计、温馨提示语定时显示程序设计和美景欣赏网站设计。

项目目标

【知识目标】掌握 page 指令、include 指令、taglib 指令，以及 include 动作元素、forward 请求转发元素。

【能力目标】能运用 JSP 脚本元素进行简单程序的编写，能灵活运用 JSP 指令、动作元素及请求转发元素。

【素质目标】培养学生的团队合作精神和精益求精的工作态度。

知识导图

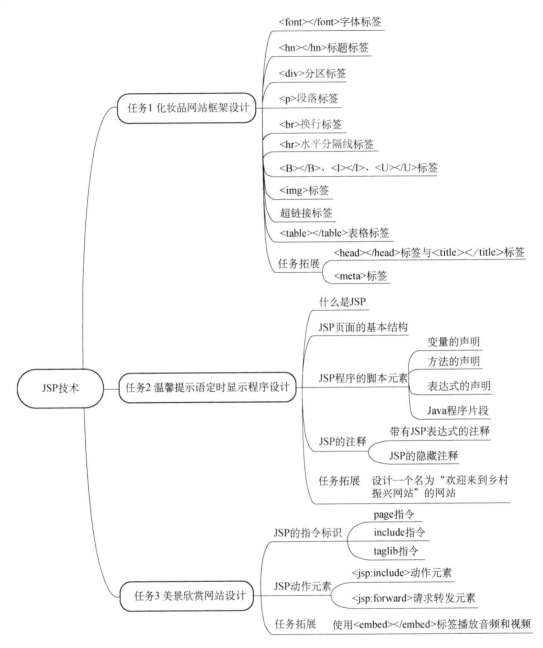

任务 1　化妆品网站框架设计

任务演示

教学视频

当我们在网上购买化妆品时，需要先在网上完成注册，然后根据注册的用户名和密码进行登录，最后浏览化妆品、选择需要的化妆品。本任务将设计一个包括导航条的化妆品网站框架,该导航条由"注册""登录""浏览化妆品""查看订单"等菜单组成。化妆品网站界面如图 2-1 所示。

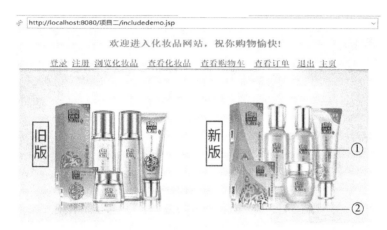

图 2-1 化妆品网站界面

知识准备

1. <font></font>字体标签

<font></font>字体标签可用于规定文本的字号、颜色、字体。

基本格式:<font[size=][color=][face=]></font>。

参数说明:

size:表示字号,共有 1~7 七种字号。数字越大,则显示的文字越大。

color:表示文字的颜色,可表示为#RRGGBB,其中,RR、GG、BB 分别表示红、绿、蓝三种颜色,可取 0~255 之间的整数,整数代表 256 种颜色强度,整数越大则颜色强度越大。由于每种颜色都有 256 种颜色强度,因此三种颜色混合后共有 256×256×256 种不同的颜色,常用的颜色名称如表 2-1 所示。

face:表示字体,中文字体有"宋体""楷体""隶书"等,西文字体有"Times New Roman""Arial""Bookman Old Style"等。

表 2-1 常用的颜色名称

中文名称	英文名称	中文名称	英文名称
粉红	Pink	纯黄	Yellow
纯蓝	Blue	金色	Gold
纯绿	Green	橙色	Orange

【例 2-1】在网页上显示"我是字体标签",并修改字体为黑体,颜色为红色,参考代码如下,运行结果如图 2-2 所示。

```
<%@ page language="java" contentType="text/html;charset=UTF-8"
pageEncoding="UTF-8"%>
<!DOCTYPE html PUBLIC "-//W3C//DTD HTML 4.01 Transitional//EN"
"http://www.w3.org/TR/html4/loose.dtd">
```

教学视频

```html
<html>
<head>
<meta http-equiv="Content-Type" content="text/html;charset=UTF-8">
<title>Insert title here</title>
</head>
<body>
<font color="red" size=40 face="黑体">我是字体标签</font>
</body>
</html>
```

图 2-2 字体标签的运行结果

2. <hn></hn>标题标签

标题是通过<h1></h1>、<h2></h2>…<h6></h6>等标签进行定义的。<h1></h1>用于定义级别最高的标题，<h6></h6>用于定义级别最低的标题。

【例 2-2】<hn></hn>标题标签的应用如下列代码所示，运行结果如图 2-3 所示。

教学视频

```html
<%@pagelanguage="java"contentType="text/html;charset=UTF-8"
pageEncoding="UTF-8"%>
<!DOCTYPEhtmlPUBLIC"-//W3C//DTD HTML 4.01 Transitional//EN""http://www.w3.org/TR/html4/loose.dtd">
<html>
<head>
<metahttp-equiv="Content-Type"content="text/html;charset=UTF-8">
<title>重庆城市职业学院</title>
</head>
<body bgcolor="f0fff0">
<h1>重庆城市职业学院</h1>
<h2>重庆城市职业学院</h2>
<h3>重庆城市职业学院</h3>
<h4>重庆城市职业学院</h4>
</body>
</html>
```

图 2-3 标题标签的运行结果

3. <div>分区标签

<div>可定义文档中的分区。<div>分区标签可以把文档分割为独立的不同部分。它可以用作严格的组织工具,并且不需要使用任何格式与其关联。如果用 id 或 class 来标记<div>分区标签,那么该标签会变得更有效。

4. <p>段落标签

<p>段落标签用于定义段落,其常用的段落标签属性 align 用于规定段落的对齐方式。

5.
换行标签

使用
换行标签可插入一个简单的换行符。

6. <hr>水平分隔线标签

<hr>水平分隔线标签可在 HTML 页面中创建一条水平分隔线,水平分隔线可以在视觉上将文档分隔成多个部分。

<hr>水平分隔线标签可选的常用属性有如下几个。
- align 属性:设置对齐方式,left、center、right 分别表示左对齐、居中对齐和右对齐。
- width 属性:设置水平分隔线的长度,以像素或百分比为单位。注意:除非 width 属性设置为小于 100%的数,否则 align 属性不会产生任何效果。
- size 属性:规定水平分隔线的高度,以像素为单位。
- color:设置水平分隔线的颜色,无此项设置则表示有阴影。
- noshade 属性:规定水平分隔线的颜色为纯色,而不是有阴影的颜色。

7. <B></B>、<I></I>、<U></U>标签

<B></B>标签用于规定粗体文本。
<I></I>标签用于显示斜体文本效果。
<U></U>标签可定义下画线文本。

教学视频

【例 2-3】综合标签应用代码如下,代码运行结果如图 2-4 所示。

```
<%@pagelanguage="java"contentType="text/html;charset=UTF-8"
pageEncoding="UTF-8"%>
<!DOCTYPEhtmlPUBLIC"-//W3C//DTD HTML 4.01 Transitional//EN""http://www.w3.org/TR/html4/loose.dtd">
<html>
<head>
<metahttp-equiv="Content-Type"content="text/html;charset=UTF-8">
<title>病毒感染</title>
</head>
<body bgcolor="yellow"text="blue">
<p align="center">针对病毒感染进行自我排查</p>
<hr size="2"width=80%>
<p align="left">针对病毒感染进行自我排查</p>
<br><br><br>
<p align="right">针对病毒感染进行自我排查</p>
</body>
</html>
```

图 2-4 综合标签的代码运行结果

8. <img>标签

可使用<img>标签向网页中嵌入一张图片。

基本格式：<img src="" alt="">。

参数说明：

src：指"source"，src属性的值是图片的URL。

alt：该属性用来为图片定义一串预备好的、可替换的文本。

【例2-4】使用<img>标签嵌入一张图片，并修改图片的宽度、高度和边框属性。

教学视频

```
<%@pagelanguage="java"contentType="text/html;charset=UTF-8"
pageEncoding="UTF-8"%>
<!DOCTYPEhtmlPUBLIC"-//W3C//DTD HTML 4.01 Transitional//EN""http://www.w3.org/
TR/html4/loose.dtd">
<html>
<head>
<metahttp-equiv="Content-Type"content="text/html;charset=UTF-8">
<title>Insert title here</title>
</head>
<body>
<img alt="图片不存在" src=" image/nurse1.JPG">
</body>
</html>
```

注意：当指定路径下的图片不存在时，显示结果如图2-5所示。若图片存在，则显示结果如图2-6所示。

图 2-5 图片不存在的显示结果

图 2-6 图片存在的显示结果

9. 超链接标签

HTML使用超链接与网络上的另一个网页相连，单击超链接则可以从一个网页跳转到另一个网页。

基本格式：<a href="url">Link text</a>。

参数说明：

href：设置目标地址。

target：设置跳转目标（未在基本格式中列举出来，详见下方示例代码）。

_self：链接的网页将在当前窗口中打开（未在基本格式中列举出来，详见下方示例代码）。

_blank：链接的网页将在新窗口中打开（未在基本格式中列举出来，详见下方示例代码）。

开始标签和结束标签之间的文字将被作为超链接显示。

【例2-5】超链接标签示例代码如下，超链接的显示结果、链接网页跳转成功的显示结果分别如图2-7、图2-8所示。

教学视频

```
<%@pagelanguage="java"contentType="text/html;charset=UTF-8"
pageEncoding="UTF-8"%>
<!DOCTYPEhtmlPUBLIC"-//W3C//DTD HTML 4.01 Transitional//EN""http://www.w3.org/
TR/html4/loose.dtd">
<html>
<head>
<metahttp-equiv="Content-Type"content="text/html;charset=UTF-8">
<title>Insert title here</title>
</head>
<body>
<a href="http://www.baidu.com">百度</a><br>
<a href="http://www.baidu.com"target="_self">百度</a><br>
<a href="http://www.baidu.com" target="_blank">百度</a><br>
</body>
</html>
```

图2-7　超链接的显示结果　　　　　图2-8　链接网页跳转成功的显示结果

10. <table></table>表格标签

表格由<table></table>标签来定义，每个表格均有若干行（由<tr></tr>标签定义），每行被分割为若干单元格（由<td></td>标签定义），其中"td"指表格数据（Table Data），即数据单元格的内容。数据单元格包含文本、图片、列表、段落、表单、水平线、表格等。

【例2-6】表格标签的示例代码如下，表格显示结果如图2-9所示。

教学视频

```
<%@page language="java"contentType="text/html;charset=UTF-8"
pageEncoding="UTF-8"%>
```

```
<!DOCTYPEhtmlPUBLIC"-//W3C//DTD HTML 4.01 Transitional//EN""http://www.w3.org/TR/html4/loose.dtd">
<html>
<head>
<metahttp-equiv="Content-Type"content="text/html;charset=UTF-8">
<title>Insert title here</title>
</head>
<bodybgcolor="f0fff0">
<tableborder="1">
<tr>
<td>编号</td>
<td>姓名</td>
<td>性别</td>
<td>工作单位</td>
</tr>
<tr>
<td> 01</td>
<td>廖丽</td>
<td>女</td>
<td>重庆城市职业学院</td>
</tr>
<td> 02</td>
<td>刘强</td>
<td>男</td>
<td>重庆科技有限公司</td>
</tr>
</table>
</body>
</html>
```

图 2-9　表格显示结果

任务实施

化妆品网站框架设计的代码如下，具体操作可扫描侧方二维码查看。

教学视频

```
<%@pagelanguage="java"contentType="text/html;charset=UTF-8"
pageEncoding="UTF-8"%>
<!DOCTYPEhtmlPUBLIC"-//W3C//DTD HTML 4.01 Transitional//EN""http://www.w3.org/TR/html4/loose.dtd">
<html>
<head>
<metahttp-equiv="Content-Type"content="text/html;charset=UTF-8">
<title>Insert title here</title>
</head>
```

```html
<body bgcolor="f0fff0">
<div align="center">
<font color="019858"><h3>欢迎进入化妆品网站,祝你购物愉快!</h3></font>
<table cellSpacing="1" cellPadding="1" width="660" align="center" border="0">
<tr valign="bottom">
<td><a href="">登录</a></td>
<td><a href="">注册</a></td>
<td><a href="">浏览化妆品</a></td>
<td><a href="">查看化妆品</a></td>
<td><a href="">查看购物车</a></td>
<td><a href="">查看订单</a></td>
<td><a href="">退出</a></td>
<td><a href="">主页</a></td>
</tr>
</table>
<img alt=" " src="image/ hzp.JPG">
</body>
</html>
```

任务拓展

1. \<head>\</head>标签与\<title>\</title>标签

\<head>\</head>标签是所有头部元素的容器。\<head>\</head>标签内的元素可包含脚本,用以指明浏览器在何处可以找到样式表、提供元信息等。可以添加到\<head>\</head>标签中的标签有\<title>\</title>、\<link>、\<meta>、\<script>\</script>、\<style>\</style>。

\<title>\</title>标签用于定义文档的标题,主要功能包括定义浏览器工具栏中的标题、提供页面被添加到收藏夹时显示的标题、显示搜索引擎的搜索结果的页面标题。

【例2-7】使用\<head>\</head>标签和\<title>\</title>标签显示七大传统节日:春节、元宵节、清明节、端午节、七夕节、中秋节、重阳节,代码如下,页面显示结果如图2-10所示。

```html
<!DOCTYPE html>
<html>
<head>
<meta charset="UTF-8">
<title>七大传统节日</title>
</head>
<body bgcolor="aaffee">
春节
元宵元
清明节
端午节
七夕节
中秋节
重阳节
</body>
</html>
```

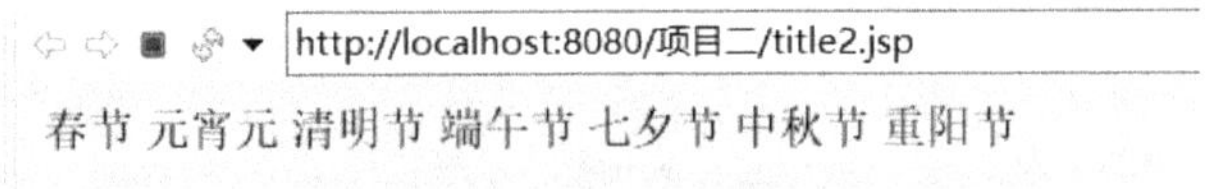

> http://localhost:8080/项目二/title2.jsp
>
> 春节 元宵元 清明节 端午节 七夕节 中秋节 重阳节

图 2-10　页面显示结果

2. <meta>标签

<meta>标签用于提供 HTML 文档的元数据。它的元素被用于规定页面的描述、关键字、文档的作者、最后的修改时间及其他元数据。<meta>标签始终位于头部元素中，<meta>标签的使用格式如下。

```
<meta 可选属性名="…" content=""/>
```

式中的 content 是<meta>标签的必选属性，其作用是描述网页的内容，一般与<meta>标签的可选属性 http-equiv 和 name 配合使用，用于指定网页的元数据信息。可将 http-equiv 可选属性与 content 必选属性组合来指定网页的编码方式，代码如下。

```
<meta http-equiv="content-type" content="text/html";charset="utf-8">
```

在 HTML5 中设置编码方式的代码如下。

```
<meta charset="UTF-8">
```

以上两行代码是等效的，建议使用时选用比较短的代码。http-equiv 可选属性与 content 必选属性组合后可以用于自动刷新网页，代码如下。

```
<meta http-equiv="Refresh" content="5">
```

此代码将使当前网页每隔 5 秒刷新一次，一般用于网址变更的情况，即将用户重定向到另外一个网址。

【例 2-8】进入 head2.html 页面 5 秒后，自动跳转到 head3.html 页面，代码如下，head2.html、head3.html 页面分别如图 2-11 和图 2-12 所示。

```
head2.html
<!DOCTYPE html>
<html>
<head>
<meta charset="UTF-8">
<title>中秋节</title>
<meta http-equiv="Refresh" content="5;url=http://localhost:8080/web/head21.html" />
</head>
<body bgcolor="aaffee">
<p>对不起，我们已经搬家了，新的地址是<a href=http://localhost:8080/项目二/head3.jsp>http://localhost:8080/项目二/head3.jsp</a>
<p>你将在 5 秒后被重新定向到新地址
</body></html>
head3.html
<!DOCTYPE html>
<html>
<head>
<meta charset="UTF-8">
```

```
<title>Insert title here</title>
</head>
<body bgcolor="aaffee">
暑退九霄净,秋澄万景清。农历八月十五,民间称为中秋,中秋赏月,品尝月饼,视为家人团圆的象征。中秋月明圆,温情满人间
</body>
</html>
```

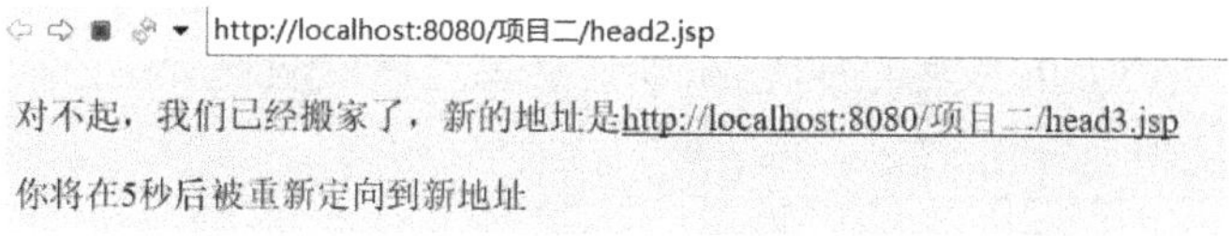

图 2-11　head2.html 页面

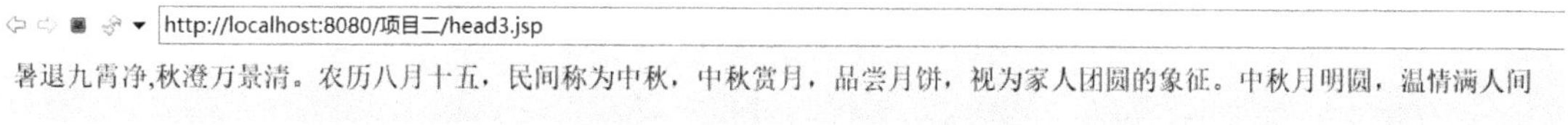

图 2-12　head3.html 页面

网页的标题、关键字对网页搜索引擎的收录和排名有影响。<meta>标签的 name 可选属性与 content 必选属性组合后,可以用来指定作者、关键字和描述等信息,如果将 name 可选属性的值设置为 "keywords",则可以在 content 必选属性中设置关键字;如果将 name 可选属性的值设置为 "Description",则可以在 content 必选属性中设置描述信息。

【例 2-9】content 必选属性和 name 可选属性的应用代码如下。

```
<!DOCTYPE html>
<html>
<head>
<meta charset="UTF-8">
<title>中秋节</title>
<meta name="keywords" content="祭月节、月光诞、月夕、秋节、仲秋节、拜月节、月娘节、月亮节、团圆节"/>
<meta name="Description" content="2006年5月20日,国务院公布首批国家级非物质文化遗产名录。自2008年起中秋节被列为国家法定节假日。" />
</head>
<body bgcolor="aaffee">
</body>
</html>
```

任务 2　温馨提示语定时显示程序设计

任务演示

设计一个程序,在一天的各时间段定时显示不同的温馨提示语,如表 2-2 所示,显示效果如图 2-13 所示。

教学视频

表 2-2　时间段与温馨提示语

时间段	温馨提示语
6:00-8:00	早上好！吃个营养丰盛的早餐
8:00-12:00	早上上班，美好开始
12:00-14:00	把工作暂时放在一边，先对自己的努力给予肯定，然后吃个营养美味的午餐吧
14:00-18:00	在这样一个不错的午后，先给自己的心情加点阳光，然后愉悦地让今天继续
18:00-21:00	下午下班，轻松过关
21:00-24:00	已经是深夜，注意休息
24:00-6:00	时间还早，再睡会吧

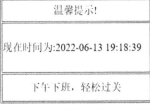

图 2-13　温馨提示语显示效果

（注：显示的时间以代码的运行时刻为准，下同）

 知识准备

1. 什么是 JSP

JSP（Java Server Pages）是由 Sun Microsystems 公司创建的一种动态网页技术标准。JSP 部署于网络服务器上，可以响应客户端发送的请求，并根据请求内容动态生成 HTML、XML 或其他格式的 Web 网页，返回给请求者。JSP 技术以 Java 语言作为脚本语言，为用户的 HTTP 请求提供服务，并能与服务器上的其他 Java 程序共同处理复杂的业务。

2. JSP 页面的基本结构

我们已经创建了 JSP 页面，但是并未对 JSP 页面的基本结构进行详细介绍，下面将详细介绍 JSP 页面的基本结构。一个 JSP 页面可以包含指令标识、HTML 代码、JavaScript 代码，以及嵌入的 Java 代码、注释和 JSP 动作标识等内容。

教学视频

【例 2-10】JSP 页面的基本结构应用代码如下，运行结果如图 2-14 所示。

```
1    <%@ page language="java" contentType="text/html;charset=UTF- 8"pageEncoding="UTF-8"%>
2    <%@ page import="java.util.Date" %>
3    <%@ page import="java.text.SimpleDateFormat" %>
4    <!DOCTYPE html PUBLIC "-//W3C//DTD HTML 4.01 Transitional//EN" "http://www.w3.org/TR/html4/loose.dtd">
5    <html>
6    <head>
7    <meta http-equiv="Content-Type" content="text/html;charset=UTF-8">
```

```
 8    <title>Insert title here</title>
 9    </head>
10    <body bgcolor="f0fff0">
11    <center>
12    <%String today=new Date().toLocaleString();%>
13    今天是:<%=today%>
14    </body>
15    </html>
```

代码解析：第 2~3 行是指令标识；第 5~15 是 HTML 代码；第 12~13 行是嵌入的 Java 代码。

图 2-14 JSP 页面的基本结构的运行结果

3. JSP 程序的脚本元素

JSP 程序主要由脚本元素组成。JSP 规范描述了 JSP 程序的三种脚本元素：声明、表达式和脚本程序。其中，声明用于声明一个或多个变量；表达式是一个完整的语言表达式；脚本程序是程序片断。所有的脚本元素都以"<%"标记开始，以"%>"标记结束。声明和表达式通过在"<%"后面加一个特殊字符进行区别。在运行 JSP 程序时，服务器可以将脚本元素转换为等效的 Java 源代码，并在服务器端执行 Java 源代码。

（1）变量的声明

在"<%!"和"%>"之间声明变量时，变量的类型可以是 Java 语言允许的任何数据类型，这些变量被称为 JSP 页面的成员变量。在声明变量时，可以一次性声明多个变量，只要声明是合法的，并且以";"结尾就可以。

【例 2-11】简单计数器的实现代码如下，运行结果如图 2-15 所示。

教学视频

```
 1  <%@ page language="java" contentType="text/html;charset=UTF-8" pageEncoding="UTF-8"%>
 2  <!DOCTYPE html PUBLIC "-//W3C//DTD HTML 4.01 Transitional//EN" "http://www.w3.org/TR/html4/loose.dtd">
 3  <html>
 4  <head>
 5  <meta http-equiv="Content-Type" content="text/html;charset=UTF-8">
 6  <title>Insert title here</title>
 7  </head>
 8  <body bgcolor="aaffee">
 9  <%! int i=0;%>
10  <p>您是第<%=++i %>个访问本站的用户</p>
11  </body>
12  </html>
```

代码解析：第 8 行的 bgcolor 用于设置背景颜色；第 9 行表示声明整型变量 i，并赋初值为 0；第 10 行先使 i 自加，然后应用表达式输出 i 的值。

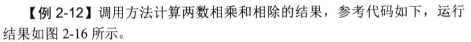

图 2-15 实现简单的计数器的运行结果

（2）方法的声明

在"<%!"和"%>"之间定义方法时，所定义的方法在整个 JSP 页面有效，并且可以在 Java 程序片段中被调用。在方法内声明的变量只在该方法内有效，当方法被调用时，在方法内声明的变量将会被分配内存，方法被调用完毕即释放这些变量所占的内存。

教学视频

【例 2-12】调用方法计算两数相乘和相除的结果，参考代码如下，运行结果如图 2-16 所示。

```
1   <%@ page language="java" contentType="text/html;charset=UTF-8"
2   pageEncoding="UTF-8"%>
3   <!DOCTYPE html PUBLIC "-//W3C//DTD HTML 4.01 Transitional//EN" "http://www.w3.org/TR/html4/loose.dtd">
4   <html>
5   <head>
6   <meta http-equiv="Content-Type" content="text/html;charset=UTF-8">
7   <title>Insert title here</title>
8   </head>
9   <%! double multi(double x,double y)
10  { return x*y;}
11  double div(double x,double y)
12  { return x/y;}  %>
13  <body bgcolor="aaffee">
14  <% double x=10,y=3;
15  out.print("x 与 y 相乘:"+multi(x,y));
16  out.println("<br>");
17  out.print("x 与 y 相除:"+div(x,y));
18  %>
19  </body>
20  </html>
```

代码解析：第 9～12 行在"<%!"和"%>"之间定义了两个方法，即 multi(double x, double y) 和 div(double x, double y)；第 13 行通过 bgcolor 设置背景颜色的颜色代码为"aaffee"；第 14～17 行在程序片段中调用了两个方法。

图 2-16 计算两数相乘和相除的运行结果

（3）表达式的声明

表达式用于在 JSP 请求处理阶段进行运算，运算结果将会转换成字符串，并与模板数据组合在一起。表达式在页面的位置就是该表达式运算结果显示的位置。

【例 2-13】在 JSP 页面中通过 JSP 表达式输出 "保护环境，爱护地球！"，参考代码如下，运行结果如图 2-17 所示。

```
<%@ page language="java" contentType="text/html;charset=UTF-8"
    pageEncoding="UTF-8"%>
<!DOCTYPE html PUBLIC "-//W3C//DTD HTML 4.01 Transitional//EN" "http://www.w3.org/TR/html4/loose.dtd">
<html>
<head>
<meta http-equiv="Content-Type" content="text/html;charset=UTF-8">
<title>Insert title here</title>
</head>
<body bgcolor="aaffee">
<%
String str="保护环境,爱护地球!";
%>
<%=str %>
</body>
</html>
```

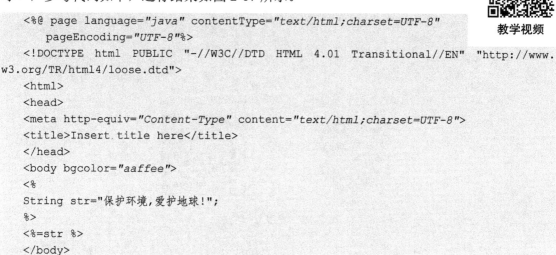

图 2-17 通过 JSP 表达式输出 "保护环境，爱护地球！" 的运行结果

（4）Java 程序片段

一个 JSP 页面可以有多个 Java 程序片段，这些 Java 程序片段将插入 "<%" 和 "%>" 之间，并被 JSP 引擎按顺序执行。在 Java 程序片段中声明的变量被称为 JSP 页面的局部变量。在多个用户请求一个 JSP 页面时，Java 程序片段将被多次执行，且分别在不同的线程中被执行。

【例 2-14】获得一个位于 7~19 之间的随机数，如果获得的数小于或等于 13，就显示一张重庆小面的图片，否则显示一张红烧牛肉面的图片，代码如下，程序运行效果如图 2-18 所示。

```
<%@ page language="java" contentType="text/html;charset=UTF-8"
    pageEncoding="UTF-8"%>
<!DOCTYPE html PUBLIC "-//W3C//DTD HTML 4.01 Transitional//EN" "http://www.w3.org/TR/html4/loose.dtd">
<html>
<head>
<meta http-equiv="Content-Type" content="text/html;charset=UTF-8">
<title>Insert title here</title>
</head>
<body>
<% int number=7+(int)(Math.random()*13);
  if(number<=13)
   { %>
  <center><h2>显示重庆小面图片</h2>
```

```
    <image src='image/noodles.JPG' width=180 height=178><br>重庆小面</image></center>
<%} else {
%>
<center><h2>显示红烧牛肉面图片</h2>
<image src="image/beef.JPG" width=180 height=178><br>红烧牛肉面</image></center>
 <%}%>
</body>
</html>
```

显示红烧牛肉面图片　　　　显示重庆小面图片

红烧牛肉面　　　　　　　重庆小面

图 2-18　程序运行效果

显示图片需要在 Java 程序片段间插入用于显示图片的 HTML 标记，其语法格式为"<image src=图片的 URL>文字说明</image>"，且要将两张名字分别为 noodles.jpg 和 beef.jpg 的图片保存到 Web 服务目录的 image 文件夹中。

4. JSP 的注释

在 JSP 规范中，可以使用两种注释方式：带有 JSP 表达式的注释、JSP 的隐藏注释，下面分别对这两种注释方式进行讲解。

（1）带有 JSP 表达式的注释

可以在 JSP 页面中嵌入 Java 程序片段，也可以在 Java 程序片段中嵌入注释，Java 程序片段的注释方法与 Java 的注释方法相同，包含以下三种情况。

第一种：单行注释。

单行注释以 "//" 开头，后面是注释内容，其语法格式如下。

```
//注释内容
```

第二种：多行注释。

多行注释以 "/*" 开头，以 "*/" 结束，在开始标识符和结束标识符之间的内容为注释内容，其语法格式如下。

```
/* 注释内容 */
```

第三种：文档注释。

文档注释是对代码结构和功能的描述，其语法格式如下。

```
/** 提示信息
*/
```

（2）JSP 的隐藏注释

JSP 隐藏注释的作用是使注释写在 JSP 程序中，但是不会发送给用户，其语法格式如下。

```
<%--注释内容--%>
```

```
<!--注释内容 -->
```

任务实施

教学视频

任务要求：首先获取当前系统时间，然后根据当前系统时间输出对应的温馨提示语，参考代码如下。

```jsp
<%@ page language="java" contentType="text/html;charset=UTF-8"
    pageEncoding="UTF-8"%>
<%@ page import="java.util.Date,java.text.*" %>
<!DOCTYPE html PUBLIC "-//W3C//DTD HTML 4.01 Transitional//EN" "http://www.w3.org/TR/html4/loose.dtd">
<html>
<head>
<meta http-equiv="Content-Type" content="text/html;charset=UTF-8">
<title>Insert title here</title>
</head>
<body>
<%
Date nowday=new Date();
int hour=nowday.getHours();
SimpleDateFormat format=new SimpleDateFormat("yyyy-MM-dd HH:mm:ss");
String time=format.format(nowday);
%>
<center>
    <table border="1" width="300">
        <tr height="40"><td align="center">温馨提示!</td></tr>
        <tr height="90"><td>现在时间为:<%=time %></td></tr>
        <tr height="50">
        <td align="center">
          <%

            if(hour>=6&&hour<8)
             out.print("早上好!吃个营养丰盛的早餐");
            else if(hour>=8&&hour<12)
             out.print("早上上班, 美好开始");
            else if(hour>=12&&hour<=14)
             out.print("把工作暂时放在一边, 先对自己的努力给予肯定, 然后吃个营养美味的午餐吧");
            else if(hour>=14&&hour<18)
             out.print("在这样一个不错的午后, 先给自己的心情加点阳光, 然后愉悦地让今天继续");
            else if(hour>=18&&hour<21)
             out.print("下午下班, 轻松过关");
            else if(hour>=21&&hour<24)
             out.print("已经是深夜, 注意休息");
            else if(hour>=24&&hour<6)
             out.print("时间还早, 再睡会吧");
          %>
        </td>
    </table>
```

```
</center>
</body>
</html>
```

任务拓展

设计一个名为"欢迎来到乡村振兴网站"的网站

要求网站包括标题、图片、访问网站的次数及版权信息，参考代码如下，程序运行结果如图 2-19 所示。

```
<body>
<h1>欢迎来到乡村振兴网站</h1>
<hr>
<img src="image/a1.JPG" alt="" >
<%! int i=1; %>
<h2>你是第<%=i++ %>个访问此网站的用户</h2>
 Copyright &copy; 2022-6  廖丽 版权所有
</body>
```

图 2-19　程序运行结果

任务 3　美景欣赏网站设计

编写一个展示"风景图"页面和"花的世界"页面的美景欣赏网站。

任务演示

美景欣赏网站首页、"花的世界"页面、"风景图"页面分别如图 2-20、图 2-21、图 2-22 所示。

图 2-20　美景欣赏网站首页

教学视频

图 2-21 "花的世界"页面

图 2-22 "风景图"页面

知识准备

1. JSP 的指令标识

指令标识在客户端是不可见的，它会被服务器解释并执行，通过指令标识可以使服务器按照设置的指令来执行动作，以及设置在整个 JSP 页面范围内有效的属性。可以在一个指令标识中设置多个属性，这些属性会影响整个 JSP 页面。JSP 主要包含三种指令，分别是 page 指令、include 指令和 taglib 指令。

指令通常以"<%@"标记开始，以"%>"标记结束，其通用格式如下。

```
<%@ 指令名称 属性1="属性值" 属性2="属性值"...%>
```

（1）page 指令

page 指令即页面指令，其定义的属性在整个 JSP 页面范围内有效，其格式如下。

```
<%@ page 属性名1="属性值1"属性名2="属性值2"...%>
```

page 指令可以放在 JSP 页面的任意行中，但为了便于程序代码的阅读，一般将其放在文件的开始部分。page 指令有多种属性，设置属性可以影响当前的 JSP 页面，其常用属性如表 2-3 所示。

表 2-3 page 指令的常用属性

属性	属性含义
language	指定 JSP 使用的脚本语言，默认为 Java
import	指定 JSP 页面中需要导入的 Java 包列表，导入多个包需要用逗号隔开
pageEncoding	指定 JSP 页面的编码方式，默认是 ISO-8859-1。若要在页面上显示中文，则需将编码方式修改为 GB2312 或 GBK
contentType	指定 JSP 页面的 MIME 类型和字符编码，例如：HTML 格式为 text/html；纯文本格式为 text/plain；JPG 图像为 image/jpeg；GIF 图像为 image/gif
session	指定 JSP 是否内置 session 对象，如果为 True，则说明内置了 session 对象，否则没有内置 session 对象，默认为 True
buffer	设置网页输出使用的缓冲区大小，默认值是 8KB

属性	属性含义
autoFlush	指定当缓冲区满时是否自动输出缓冲区的数据。如果为 True，则输出正常，否则当缓冲区满时将抛出异常，默认为 True
info	指定一个可以在 Servlet 中通过 getServletInfo() 方法获得的字符串
errorPage	指定一个 JSP 页面，让此页面来处理当前页面抛出的但未被捕获的异常
isErrorPage	表示当前页面是否可以作用于其他错误页面

【例 2-15】实现在 JSP 页面中显示中文和当前时间，参考代码如下，运行结果如图 2-23 所示。

```
1   <%@ page language="java" contentType="text/html;charset=UTF-8"
        pageEncoding="UTF-8"%>
2   <%@ page import="java.util.Date" %>
3   <%@ page import="java.text.SimpleDateFormat" %>
4   <!DOCTYPE html PUBLIC "-//W3C//DTD HTML 4.01 Transitional//EN" "http://www.w3.org/TR/html4/loose.dtd">
5   <html>
6   <head>
7   <meta http-equiv="Content-Type" content="text/html;charset=UTF-8">
8   <title>Insert title here</title>
9   </head>
10  <body bgcolor="f0fff0">
11  <center>
12  <%String today=new Date().toLocaleString();%>
13  今天是:<%=today%>
14  </body>
15  </html>
```

http://localhost:8080/项目二/date.jsp

今天是：2022-9-1 10:14:01

图 2-23　在 JSP 页面中显示中文和当前时间的运行结果

代码解析：第 1 行的 language 属性指定了使用的脚本语言是 Java，contentType 属性指定当前页面的类型和编码方式，代码"text/html"指定当前页面的 MIME 类型，代码"charset=UTF-8"指定当前页面的编码方式为 UTF-8，可显示中文字符。第 2~3 行表示导入 Java 类需要的包。

（2）include 指令

在实际开发中，有时需要在一个 JSP 页面中包含另一个 JSP 页面，这时可以通过 include 指令实现，include 指令的具体语法格式如下。

```
<%@ include file="fileurl"
```

参数说明：file 指定静态文件的路径，其值"fileurl"可以为相对路径，也可以为绝对路径，一般为相对路径。

【例 2-16】设计两个 JSP 文件，一个为 head.jsp，用于显示框架的头部；另一个为

includedemo.jsp，它是化妆品网站的主界面，如图 2-24 所示。

教学视频

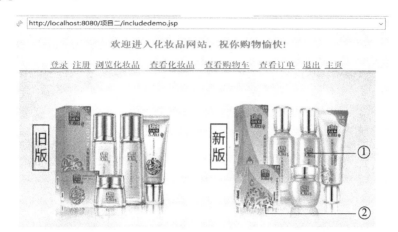

图 2-24 化妆品网站的主界面

第一步：编写 head.jsp 文件。

```jsp
<%@ page language="java" contentType="text/html;charset=UTF-8"
    pageEncoding="UTF-8"%>
<!DOCTYPE html PUBLIC "-//W3C//DTD HTML 4.01 Transitional//EN" "http://www.w3.org/TR/html4/loose.dtd">
<html>
<head>
<meta http-equiv="Content-Type" content="text/html;charset=UTF-8">
<title>Insert title here</title>
</head>
<body bgcolor="f0fff0">
<div align="center">
<table cellSpacing="1" cellPadding="1" width="660" align="center" border="0">
<tr valign="bottom">
<td><a href="">登录</a></td>
<td><a href="">注册</a></td>
<td><a href="">浏览化妆品</a></td>
<td><a href="">查看化妆品</a></td>
<td><a href="">查看购物车</a></td>
<td><a href="">查看订单</a></td>
<td><a href="">退出</a></td>
<td><a href="">主页</a></td>
</tr>
</table>
</div>
</body>
</html>
```

第二步：编写 includedemo.jsp 文件。

```jsp
<%@ page language="java" contentType="text/html;charset=UTF-8"
```

```
        pageEncoding="UTF-8"%>
<!DOCTYPE html PUBLIC "-//W3C//DTD HTML 4.01 Transitional//EN" "http://www.w3.org/TR/html4/loose.dtd">
<html>
<head>
<meta http-equiv="Content-Type" content="text/html;charset=UTF-8">
<title>Insert title here</title>
</head>
<body bgcolor="f0fff0">
<div align="center">
<h3><font color="019858">欢迎进入化妆品网站,祝你购物愉快!</font></h3>
<%@include file="head.jsp" %>
<img alt="" src="image/hzp.JPG">
</div>
</body>
</html>
```

（3）taglib 指令

taglib 指令可用于表明当前 JSP 文件使用的自定义标签,并指明自定义标签的前缀,其语法格式如下。关于 JSP 的自定义标签和 taglib 指令的使用将在项目七进行详细介绍。

```
<%@ taglib uri="URIToTagLibrary" prefix="tagPrefix"%>
```

参数说明:uri 指定自定义标签的 URL 地址,prefix 指定自定义标签的前缀。

2. JSP 动作元素

JSP 动作元素用于控制 JSP 的行为,执行一些常见的 JSP 页面动作。通过 JSP 动作元素可以实现使用多行 Java 代码才能实现的效果,如包含页面文件、实现请求转发等。常用的 JSP 动作元素有<jsp:include>、<jsp:forward>、<jsp:param>、<jsp:useBean>、<jsp:setProperty>、<jsp:getProperty>、<jsp:fallback>、<jsp:plugin>,下面介绍比较常用的两个 JSP 动作元素。

（1）<jsp:include>动作元素

<jsp:include>动作元素用于向当前页面引入其他文件,被引入的文件可以是动态文件,也可以是静态文件,其基本格式如下。

```
<jsp:include page= "url" flush= "true|false">
</jsp:include>
```

向被包含的动态页面中传递参数的基本格式如下。

```
<jsp:include page= "url" flush= "true|false">
    <jsp:param name="参数名称" value="参数值"
</jsp:include>
```

参数说明:

page 属性:该属性指定被包含的动态页面的路径,该路径可以为相对路径,也可以为绝对路径。

flush 属性:表示当输出的缓冲区满时是否清空缓冲区。该属性值为 boolean 类型,默认值为 False,通常情况下需要改为 True。

<jsp:param>子标识：可以向被包含的动态页面传递参数。

<jsp:include>标识：对被包含的动态页面和静态页面的处理方式是不同的。如果被包含的是静态页面，则在执行静态页面后，使用该标识的位置处将会输出这个静态页面的内容。如果被包含的是动态页面，那么 JSP 编译器将编译并执行这个动态页面。不能通过页面的名称来判断该页面是静态的还是动态的，需要使用<jsp:include>标识识别页面的类型。

【例 2-17】设计一个用户登录界面，实现用户名、密码和验证码的输入，如图 2-25 所示。

分析：设计两个 JSP 页面，一个 JSP 页面用于产生随机的由四位数构成的验证码，如图 2-26 所示，另一个 JSP 页面是登录界面，包含验证码页码。下面分两步完成。

教学视频

第一步：随机产生一个四位数并显示在页面上，参考代码如下。

图 2-25　用户登录界面　　　　　　　　　图 2-26　验证码

```
<%@ page language="java" contentType="text/html;charset=UTF-8" pageEncoding="UTF-8"%>
<%@ page import="java.util.*" %>
<!DOCTYPE html PUBLIC "-//W3C//DTD HTML 4.01 Transitional//EN" "http://www.w3.org/TR/html4/loose.dtd">
<html>
<head>
<meta http-equiv="Content-Type" content="text/html;charset=UTF-8">
<title>Insert title here</title>
</head>
<body bgcolor="f0fff0">
<%
Random rand=new Random();
  for(int i=0;i<4;i++)
  {
  int start=rand.nextInt(10);
  out.print(" "+start);
  }
%>
</body>
</html>
```

第二步：设计用户登录界面。

```
<%@ page language="java" contentType="text/html;charset=UTF-8" pageEncoding="UTF-8"%>
<!DOCTYPE html PUBLIC "-//W3C//DTD HTML 4.01 Transitional//EN" "http://www.w3.org/TR/html4/loose.dtd">
<html>
```

```
<head>
<meta http-equiv="Content-Type" content="text/html;charset=UTF-8">
<title>Insert title here</title>
</head>
<body>
<center>
<form action=" " method="post">
 <table>
  <tr>
    <td>用户名</td>
    <td><input type="text" name="user"></td>
  </tr>
  <tr>
    <td>密  码</td>
    <td><input type="password" name="pwd"></td>
  </tr>
<tr>
   <td>验证码</td>
   <td><input type="text" name="confire"></td>
    <td> <jsp:include page="Verification.jsp"></jsp:include>
  </tr>
  <tr align="center">
  <td><input type="submit" value="提交"></td>
  <td><input type="reset" value="重置"> </td>
  </tr>
  </table>
</form>
</center>
</body>
</html>
```

【例 2-18】 计算三角形的面积，并显示在页面上。

要求：通过 sides.jsp 页面计算并显示三角形面积，当 triangle.jsp 被加载时会获取 sides.jsp 页面上的 include 标识的 param 子标识提供的三条边的长度，运行结果如图 2-27 所示。

教学视频

第一步：新建一个 sides.jsp 页面。

```
<body>
<%
double a=3,b=4,c=5;
%>
<br>计算三角形的面积的三条边为<%=a %>,<%=b %>,<%=c %>
<jsp:include page="triangle.jsp">
<jsp:param value="<%=a %>" name="sidea"/>
<jsp:param value="<%=b %>" name="sideb"/>
<jsp:param value="<%=c %>" name="sidec"/>
</jsp:include>
</body>
```

第二步：新建一个 triangle.jsp 页面。

```jsp
<%! public String getArea(double a,double b,double c)
{   if(a+b>c&&a+c>b&&b+c>a)
{   double p=(a+b+c)/2;
double area=Math.sqrt(p*(p-a)*(p-b)*(p-c));
return ""+area; } else
{return "不能构成三角形,无法计算面积";}}%>

<body bgcolor="f0fff0">
<%String a=request.getParameter("sidea");
String b=request.getParameter("sideb");
String c=request.getParameter("sidec");
double sidea=Double.parseDouble(a);
double sideb=Double.parseDouble(b);
double sidec=Double.parseDouble(c);
%>
<font size=4 color=blue>
<br>传来三条边的值是:<%=sidea %>,<%=sideb %>,<%=sidec %><br>
三角形的面积是:<%=getArea(sidea,sideb,sidec) %><br></font>
</body>
```

http://localhost:8080/项目二/sides.jsp

计算三角形的面积的三条边为3.0,4.0,5.0
传来三条边的值是：3.0,4.0,5.0
三角形的面积是：6.0

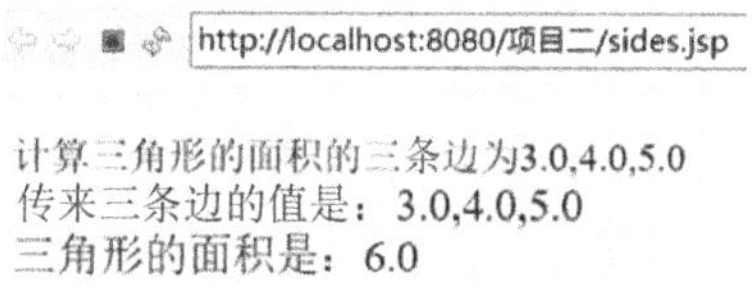

图 2-27 运行结果

（2）<jsp:forward>请求转发元素

<jsp:forward>请求转发元素可以将当前请求转发到其他 Web 资源，如 HTML 页面、JSP 页面和 Servlet 等。请求被转发之后，当前页面将不再被执行，而是执行该请求转发元素指定的目标页面，基本格式如下。

```
<jsp:forward page="要转向的页面" />
```

也可以写成如下的形式。

```
<jsp:forward page="要转向的页面" >
    param 子标记
    </jsp:forward>
```

<jsp:forward>请求转发元素用来转发用户的请求，使用户从请求的页面跳转到另一个页面，此跳转为服务器端的跳转，用户的地址栏不会产生变化。"forward"之前的代码会被执行，而其之后的代码不会被执行。

【例 2-19】<jsp:forward>请求转发元素的应用代码如下，程序运行结果如图 2-28 所示。

第一步，编写 forword1.jsp 代码。

```
<%@ page language="java" contentType="text/html;charset=UTF-8"
```

```
    pageEncoding="UTF-8"%>
<!DOCTYPE html PUBLIC "-//W3C//DTD HTML 4.01 Transitional//EN" "http://www.w3.org/TR/html4/loose.dtd">
<html>
<head>
<meta http-equiv="Content-Type" content="text/html;charset=UTF-8">
<title>Insert title here</title>
</head>
<body bgcolor="f0fff0">
<jsp:forward page="news.jsp"></jsp:forward>
<% out.print("此项代码将不会被执行");%>
</body>
</html>
```

第二步:编写 news.jsp 代码。

```
<%@ page language="java" contentType="text/html;charset=UTF-8"
    pageEncoding="UTF-8"%>
<!DOCTYPE html PUBLIC "-//W3C//DTD HTML 4.01 Transitional//EN" "http://www.w3.org/TR/html4/loose.dtd">
<html>
<head>
<meta http-equiv="Content-Type" content="text/html;charset=UTF-8">
<title>Insert title here</title>
</head>
<body bgcolor="f0fff0">
<div>
北京时间 2022 年 6 月 5 日 10 时 44 分,据中国载人航天工程办公室消息,搭载神舟十四号载人飞船的长征二号 F 遥十四运载火箭在酒泉卫星发射中心点火发射,约 577 秒后,神舟十四号载人飞船与火箭成功分离,进入预定轨道,飞行乘组状态良好,发射取得圆满成功。
</div>
</body>
</html>
```

图 2-28　程序运行结果

任务实施

任务要求:

① 编写四个 JSP 页面:one.jsp、two.jsp、three.jsp 和 error1.jsp。one.jsp、two.jsp、three.jsp 页面都包含一个导航条,可以通过超链接访问这三个页面。

② one.jsp、two.jsp、three.jsp 都通过 include 指令动态加载导航条文件 head1.jsp。

③ 根据在 one.jsp 页面输入的数据判断执行哪个页面,如果输入的是 1~300 的整数,则执行"风景图"页面,如果输入的是 301~400 的整数,则执行"花的世界"页面,否则

跳转到 error1.jsp 页面。

第一步：创建 head1.jsp 页面。

```jsp
<%@ page language="java" contentType="text/html;charset=UTF-8"
    pageEncoding="UTF-8"%>
<!DOCTYPE html PUBLIC "-//W3C//DTD HTML 4.01 Transitional//EN" "http://www.w3.org/TR/html4/loose.dtd">
<html>
<head>
<meta http-equiv="Content-Type" content="text/html;charset=UTF-8">
<title>Insert title here</title>
</head>
<body>
<table cellspacing="1" cellpadding="1" width="60%" align="center" border="0">
<tr valign="bottom">
<td><a href="one.jsp">首页</a></td>
<td><a href="two.jsp">风景图</a></td>
<td><a href="three.jsp">花的世界</a></td>
</tr>
</table>
</body>
</html>
```

第二步：创建 one.jsp 页面。

```jsp
<%@ page language="java" contentType="text/html;charset=UTF-8"
    pageEncoding="UTF-8"%>
<!DOCTYPE html PUBLIC "-//W3C//DTD HTML 4.01 Transitional//EN" "http://www.w3.org/TR/html4/loose.dtd">
<html>
<head>
<meta http-equiv="Content-Type" content="text/html;charset=UTF-8">
<title>Insert title here</title>
</head>
<body bgcolor="f0fff0">
<jsp:include page="head1.jsp"></jsp:include>
<form action="" method="get" name="form">
请输入 1~400 的整数:
<input type="text" name="number"> <br/>
<input type="submit" value="送出" name="submit"/>
</form>
<%
String num=request.getParameter("number");
if(num==null)
    num="0";
try{
    int n=Integer.parseInt(num);
    if(n>=1&&n<=300)
    {
```

```
%>
<jsp:forward page="two.jsp">
<jsp:param value="<%=n %>" name="number"/>
</jsp:forward>
<%} else if(n>300&&n<=400) {%>
<jsp:forward page="three.jsp">
<jsp:param value="<%=n %>" name="number"/>
</jsp:forward>
<% }else if(n>400){ %>
<jsp:forward page="error.jsp">
  <jsp:param value="<%=n %>" name="mess"/>
</jsp:forward>
<%} %>
<%}catch(Exception e){
%>
 <jsp:forward page="error.jsp">
  <jsp:param value="<%=e.toString() %>" name="mess"/>
 </jsp:forward>
 <%} %>
</body>
</html>
```

第三步：创建 two.jsp 页面。

```
<%@ page language="java" contentType="text/html;charset=UTF-8"
    pageEncoding="UTF-8"%>
<!DOCTYPE html PUBLIC "-//W3C//DTD HTML 4.01 Transitional//EN" "http://www.w3.org/TR/html4/loose.dtd">
<html>
<head>
<meta http-equiv="Content-Type" content="text/html;charset=UTF-8">
<title>Insert title here</title>
</head>
<body bgcolor="f0fff0">
<div align="center">
<jsp:include page="head1.jsp"></jsp:include>
<h1>风景图</h1>
<%
String s=request.getParameter("number");
%>
<h2>瀑布</h2>
<img alt="" src="image/pb.JPG" width="<%=s %>" height="<%=s %>" />
<p>日照香炉生紫烟,遥看瀑布挂前川;<br>
飞流直下三千尺,疑是银河落九天。
</p>
<h2>登高</h2>
<img alt="" src="image/cj.JPG" width="<%=s %>" height="<%=s %>" />
<p>风急天高猿啸哀,渚清沙白鸟飞回;<br>
```

```
无边落木萧萧下,不尽长江滚滚来。</p>
</div>
</body>
</html>
```

第四步:创建 three.jsp 页面。

```
<%@ page language="java" contentType="text/html;charset=UTF-8"
    pageEncoding="UTF-8"%>
<!DOCTYPE html PUBLIC "-//W3C//DTD HTML 4.01 Transitional//EN" "http://www.w3.org/TR/html4/loose.dtd">
<html>
<head>
<meta http-equiv="Content-Type" content="text/html;charset=UTF-8">
<title>Insert title here</title>
</head>
<body bgcolor="f0fff0">
<div align="center">
<jsp:include page="head1.jsp"></jsp:include>
<h1>花的世界</h1>
<%
String s=request.getParameter("number");
%>
<h2>郁金香</h2>
<img alt="" src="image/yjx.JPG" width="<%=s %>" height="<%=s %>" />
<p>兰陵美酒郁金香,玉碗盛来琥珀光;<br/>
但使主人能醉客,不知何处是他乡。
</p>
<h2>朱顶红</h2>
<img alt="" src="image/zdh.JPG" width="<%=s %>" height="<%=s %>" />
<p>一柱擎天发,状似君子兰;<br/>
簇簇红似火,一心向天燃。</p>
</div>
</body>
</html>
```

第五步:创建 error1.jsp 页面。

```
<%@ page language="java" contentType="text/html;charset=UTF-8"
    pageEncoding="UTF-8"%>
<!DOCTYPE html PUBLIC "-//W3C//DTD HTML 4.01 Transitional//EN" "http://www.w3.org/TR/html4/loose.dtd">
<html>
<head>
<meta http-equiv="Content-Type" content="text/html;charset=UTF-8">
<title>Insert title here</title>
</head>
<body bgcolor="yellow">
<jsp:include page="head1.jsp"></jsp:include>
```

```
<h1>this is error.jsp</h1>
<%
String s=request.getParameter("mess");
out.print("<br>传递过来的错误信息"+s);
%>
<br><img alt="" src="error.JPG" width="120" height="120"/>
</body>
</html>
```

任务拓展

使用<embed></embed>标签播放音频和视频

<embed></embed>标签可以播放音频和视频，当浏览器执行该标签时，浏览器所在的设备上的默认播放器会嵌入浏览器中，以便播放音频或视频，其基本格式如下。

```
<embed src="音频或视频的url">描述文字</embed>
```

如果音频和视频与当前页面在同一 Web 服务目录中，那么<embed></embed>标签中的 src 属性的值就是该文件的名称。如果音频和视频在当前 Web 服务目录的一个子目录中，比如 avi 子目录，那么<embed></embed>标签中的 src 属性的值就是 avi 子目录中的该文件的名称。

<embed></embed>标签中常用的属性如下。

autostart 属性：属性值为 True 或 False，其属性值用来指定音频或视频在传送完毕后是否立刻播放。该属性的默认值是 False。

loop 属性：若该属性取值为正整数，则指定音频或视频重复播放的次数；若该属性取值为-1，则无限次循环播放。

width 属性和 height 属性：取值均为正整数，分别用于指定播放器的宽度和高度。如果省略 width 属性和 height 属性，则使用默认值。

下面设计一个可以播放视频的 JSP 页面，代码如下，播放视频的 JSP 页面如图 2-29 所示。

```
<%@ page language="java" contentType="text/html;charset=UTF-8" pageEncoding="UTF-8"%>
<!DOCTYPE html PUBLIC "-//W3C//DTD HTML 4.01 Transitional//EN" "http://www.w3.org/TR/html4/loose.dtd">
<html>
<head>
<meta http-equiv="Content-Type" content="text/html;charset=UTF-8">
<title>Insert title here</title>
</head>
<body>
<form action="" method=post name=form>
  选择视频:<br>
  <select id="select" name="video1">
    <option value="我的祖国.mp4" selected>我的祖国
    <option value="梦幻.mp4">梦幻
    <option value="夕阳山顶.mp4">夕阳山顶
```

```
    </select>
    <input type="submit" value="提交" name="submit">
</form>
<video id="video" src="avi/我的祖国.mp4" width="400" height="225" autoplay muted=
"true" controls controlslist="nodownload nofullscreen">
</video>
<script>
    let select = document.getElementById("select");
    let video = document.getElementById("video");
    let selectVal = select.value;
    select.addEventListener('change',() => {
      selectVal = select.value;
      video.src = "avi/" + selectVal
    })
    console.log(selectVal,video.src)
</script>
</body>
</html>
```

图 2-29 播放视频的 JSP 页面

项 目 实 训

实训一 设计中秋节网站框架

古代有不少征夫、商贾、官员、文人因为职责或生计,远离故土,因此中秋夜的圆月总能唤起人们对团圆的向往、对亲人的思念。请基于上述背景设计一个中秋节网站框架,包含"登录""视频""诗歌""音乐""美食"等菜单,如图 2-30 所示。

实训二 设计信贷数据分析可视化平台用户注册界面

信贷数据分析可视化平台用户注册界面如图 2-31 所示。

图 2-30　中秋节网站框架

图 2-31　信贷数据分析可视化平台用户注册界面

课 后 练 习

一、填空题

1. 在 JSP 的指令中，用来定义与页面相关的属性的指令是_____；用来在一个 JSP 页面中包含另一个 JSP 页面的指令是_____；用来定义一个标签库及其自定义标签前缀的指令是_____。

2. _____动作元素允许在页面被请求时包含一些其他资源，如一个静态的 HTML 文件或动态的 JSP 文件。

3. page 指令的 MIME 类型的默认值为 text/html，默认字符集为_____。

4. 指令分为三种，它们分别是_____、_____、_____。

5. _____是一段在客户端请求时需要先被服务器执行的 Java 代码，它可以产生输出内容，并把输出内容发送到客户端的输出流中，同时也可以是一段流控制语句。

7. _____动作元素允许将请求转发到其他的 HTML 文件、JSP 文件或者一个程序片段。

9. page 指令的 MIME 类型的默认值为_____，默认字符集为_____。

二、选择题

1. 下列不符合 JSP 标识符规定的是（　　）。
A. System1　　　　B. studnt_info　　　　C. Main　　　　D. 8ab_

2. 下列关于 JSP 指令的描述正确的是（　　）。
A. 指令以 "<%@" 开始，以 "%>" 结束
B. 指令以 "<%" 开始，以 "%>" 结束
C. 指令以 "<" 开始，以 ">" 结束
D. 指令以 "<jsp:" 开始，以 "/>" 结束

3. JSP 代码 "<%="6+7"%>" 将输出（　　）。
A. 6+7　　　　　　　　　　　　　　B. 13
C. 67　　　　　　　　　　　　　　　D. 表达式有错误，不输出结果

4. 下列选项中，（　　）是正确的表达式。
A. <%! Float a=3.5;%>　　　　　　B. <% float a=3.5;%>
C. <%=(4+7);%>　　　　　　　　　D. <%=(4+7)%>

5. page 指令的（　　）属性可用来引用包或类。
A. extends　　　B. import　　　C. isErrorPage　　　D. language

6. page 指令中的（　　）属性可多次出现。
A. extends　　　B. import　　　C. contentType　　　D. 不存在这样的属性

7. 以下哪些属性是 include 指令所具有的?（　　）
A. page　　　　B. file　　　　C. contentType　　　D. prefile

三、编程题

1. 在 JSP 页面中通过 JSP 表达式输出 "拼搏到无能为力，坚持到感动自我，只要心向阳光，生活就会充满期望。"
2. 应用 JSP 编写一个简单的计数器程序。
3. 编写一个 JSP 页面，并输出九九乘法表。
4. 应用 Eclipse 新建一个 Web 项目，并在该项目的根目录下创建 index.jsp 页面和 welcome.jsp 页面，要求该项目能在访问 index.jsp 页面后，自动跳转到 welcome.jsp 页面。

项目三　JSP 内置对象

项目要求

本项目是 JSP 内置对象的应用，主要实现响应客户端的请求、向客户端发送数据等功能，并应用 JSP 内置对象编写简单的应用程序。

项目分析

要完成项目任务，至少需要具备两个基本条件：一是掌握 JSP 内置对象的语法规范，二是精通 JSP 内置对象的应用。该项目分为 5 个任务，分别是应用 request 对象设计网上考试系统、应用 response 对象设计化妆品网站登录界面、应用 session 对象设计火锅点餐系统、应用 application 对象设计留言板、应用 Cookie 对象制作站点计数器。

项目目标

【知识目标】掌握<jsp:include>和<jsp:forward>的基本格式；掌握 session 对象及常用方法的应用；掌握 application 对象的基本特征及常用方法；熟悉 Cookie 对象的应用。

【能力目标】能应用 request 对象处理表单信息；能应用 response 对象响应各种信息；能应用 session 对象、application 对象实现在多个程序或用户之间共享数据；能应用 Cookie 对象精确统计站点数据。

【素质目标】培养学生良好的思考能力和分析问题的能力。

知识导图

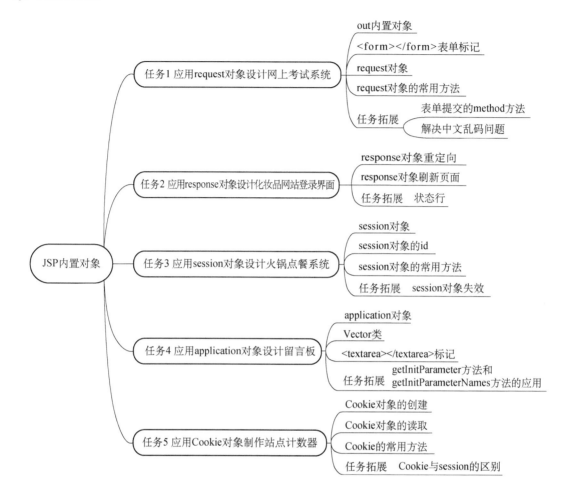

任务 1　应用 request 对象设计网上考试系统

任务演示

随着计算机时代的到来，以及 Internet 的迅速发展，计算机无处不在，它让人们的工作效率大幅提高，同时也对教育的发展形成新的推动力。考试测试作为教学中的一个重要环节，是衡量学习成果的重要手段。网上考试系统可以节省大量的人力、物力与财力，也可以大幅度增加考试的客观性和公正性。本任务是设计一个网上考试系统，考试试题页面和考试结果页面分别如图 3-1 和图 3-2 所示。

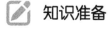

教学视频

知识准备

为了方便开发 Web 应用程序，JSP 内置了一些默认的对象，这些对象不需要预先声明就可以在脚本代码和表达式中随意使用。JSP 提供的内置对象如表 3-1 所示，所有的 JSP 代码都可以直接访问这些内置对象。

图 3-1　考试试题页面

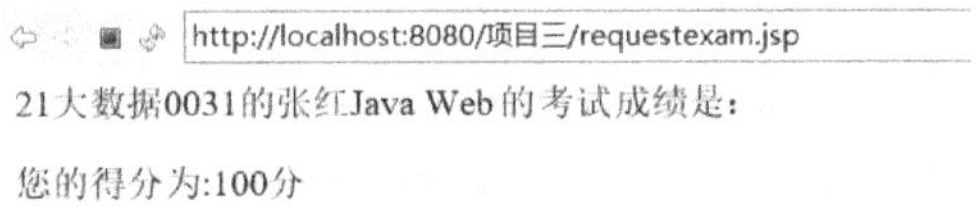

图 3-2　考试结果页面

表 3-1　JSP 提供的内置对象

内置对象名称	所属类型	有效范围	说明
out	javax.servlet.jsp.JspWriter	page	该对象提供对输出流的访问
request	javax.servlet.http.HttpServletRequest	request	该对象提供对 HTTP 请求数据的访问
response	javax.servlet.http.HttpServletResponse	page	该对象允许直接访问 HttpServletResponse 对象，可用来向客户端输出数据
session	javax.servlet.http.HttpSession	session	该对象可用来保存服务器与客户端之间的数据，当客户端关闭网站的所有网页时，session 变量会自动消失
application	javax.servlet.ServletContext	application	该对象代表应用程序的上下文，它允许 JSP 页面与包括在同一应用程序中的任何 Web 组件共享信息
config	javax.servlet.ServletConfig	page	该对象允许将初始化数据传递给 JSP 页面
exception	java.lang.Throwable	page	该对象含有只能由指定的 JSP 错误处理页面访问的异常数据
page	javax.servlet.jsp.HttpJspPage	page	该对象代表与 JSP 页面对应的 Servlet 类实例
pageContext	javax.servlet.jsp.PageContext	page	该对象是 JSP 页面的上下文，它提供了唯一一组方法来管理具有不同作用域的属性

1. out 内置对象

out 内置对象用来向客户端输出内容，被封装在 javax.servlet.jsp.JspWriter 接口中。out 内置对象是由 pageContext 对象初始化获得的，out 内置对象的作用域是 page，可使用 out 内置对象输出用户登录后的信息，out 内置对象的常用方法如表 3-2 所示。

表 3-2　out 内置对象的常用方法

方法名	说明
print()	输出各种类型的数据
println()	输出各种类型的数据并换行
newLine()	输出一个换行符

（续表）

方法名	说明
close()	关闭输出流
flush()	输出缓冲区的数据
clearBuffer()	清除缓冲区的数据，并把数据写到客户端
clear()	清除缓冲区的数据，但不把数据写到客户端
getBufferSize()	获取缓冲区的大小
isAutoFlush()	判断缓冲区是否自动刷新

【例 3-1】out 内置对象的应用代码如下，运行结果如图 3-3 所示。

```jsp
<%@ page language="java" contentType="text/html;charset=UTF-8"
    pageEncoding="UTF-8"%>
<!DOCTYPE html PUBLIC "-//W3C//DTD HTML 4.01 Transitional//EN"
"http://www.w3.org/TR/html4/loose.dtd">
<html>
<head>
<meta http-equiv="Content-Type" content="text/html;charset=UTF-8">
<title>Insert title here</title>
</head>
<body bgcolor="f0fff0">
<%
 out.print("输出字符数组<br>");
out.print(new char[]{'a','b','c'});
out.print("<br>");
out.print("输出字符");
out.print("<br>1");
out.print("<br>输出整数");
out.print(5);
out.print("<br>输出单精度数 ");
out.print(12.5f);        %>
</body>
</html>
```

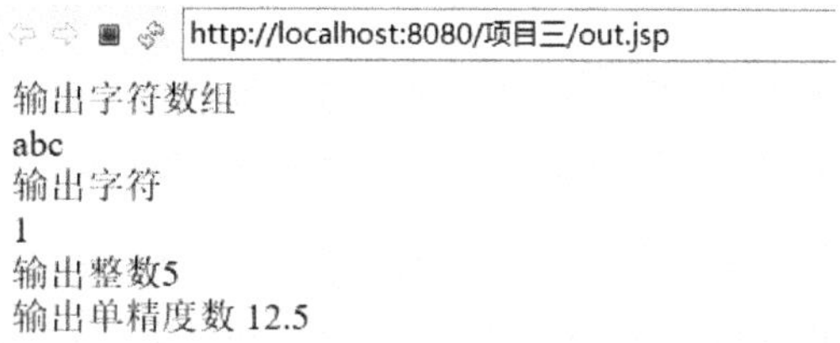

教学视频

http://localhost:8080/项目三/out.jsp

输出字符数组
abc
输出字符
1
输出整数5
输出单精度数 12.5

图 3-3 out 内置对象的运行结果

2. <form></form>表单标记

用户经常需要使用表单向服务器传输数据，<form></form>表单标记的基本格式如下。

```html
<form  action= "提交信息的目的页面"  method= get| post  name="表单的名称">
    数据提交手段部分
</form>
```

其中，<form></form>是表单标记；method属性的取值为"get"或"post"，分别代表GET方法和POST方法。GET方法和POST方法的主要区别：使用GET方法提交的信息会在提交过程中显示在浏览器的地址栏中，而使用POST方法提交的信息不会显示在浏览器的地址栏中。提交手段包括文本框、列表框等。

教学视频

【例3-2】创建一个输入信息的表单，参考代码如下，运行结果如图3-4所示。

```jsp
<%@ page language="java" contentType="text/html;charset=UTF-8"
    pageEncoding="UTF-8"%>
<!DOCTYPE html PUBLIC "-//W3C//DTD HTML 4.01 Transitional//EN" "http://www.w3.org/TR/html4/loose.dtd">
<html>
<head>
<meta http-equiv="Content-Type" content="text/html;charset=UTF-8">
<title>Insert title here</title>
</head>
<body>
    <form action="" method="post">
    <input type="text" name="str">
    <input type="submit" value="提交">
    </form>
</body>
</html>
```

图3-4 创建一个输入信息的表单的运行结果

3. request对象

request对象是一个与请求相关的HttpServletRequest类的对象。客户端先可以通过HTML表单或网页地址提供的方法提交数据，然后通过request对象的相关方法来获取这些数据。

request对象封装了客户端提交的信息，通过调用该对象的相应方法可以获取客户端提交的信息。获取表单的信息有如下四种方法。

① getParameter()：获得客户端请求的指定属性值，如果该属性值不存在，则返回null，只有属性值唯一才能使用该方法。

② getParameterNames()：获得客户端请求的所有属性，该方法的返回值的数据类型为枚举类型，如果属性不存在，则返回一个空枚举。

③ getParameterMap()：获得客户端请求的所有属性及其对应的值，返回值是Map类型的，其中键值是一个字符串，其对应的值是一个字符串数组。

④ getParameterValue()：获得指定属性的值，如果属性不存在，则返回空值，如果属性值是唯一的，则返回的数组长度为1。

【例3-3】显示例3-2输入的信息，参考代码如下，运行结果如图3-5所示。

```jsp
<%@ page language="java" contentType="text/html;charset=UTF-8"
    pageEncoding="UTF-8"%>
<!DOCTYPE html PUBLIC "-//W3C//DTD HTML 4.01 Transitional//EN" "http://www.w3.org/TR/html4/loose.dtd">
<html>
<head>
<meta http-equiv="Content-Type" content="text/html;charset=UTF-8">
<title>Insert title here</title>
</head>
<body>
<%
request.setCharacterEncoding("utf-8");
String s=request.getParameter("str");
out.print(s);
%>
</body>
</html>
```

图 3-5 显示输入的信息的运行结果

【例3-4】显示用户选择的爱好，参考代码如下，运行结果如图3-6所示。

第一步：创建 hobby.jsp 页面。

```jsp
<%@ page language="java" contentType="text/html;charset=UTF-8"
    pageEncoding="UTF-8"%>
<!DOCTYPE html PUBLIC "-//W3C//DTD HTML 4.01 Transitional//EN" "http://www.w3.org/TR/html4/loose.dtd">
<html>
<head>
<meta http-equiv="Content-Type" content="text/html;charset=UTF-8">
<title>Insert title here</title>
</head>
<body bgcolor="f0fff0">
<form action="showhobby.jsp" method="post">
用户名:<input type="text" name="username">  <br/>
密码:<input type="password" name="password">  <br/>
爱好:<input type="checkbox" name="hobby" value="sing"/>唱歌
<input type="checkbox" name="hobby" value="dangce"/>跳舞
<input type="checkbox" name="hobby" value="football"/>足球<br/>
<input type="submit" value="提交"/>
</form>
</html>
```

教学视频

第二步：创建 showhobby.jsp 页面。

```jsp
<%@ page language="java" contentType="text/html;charset=UTF-8"
```

```
    pageEncoding="UTF-8"%>
<!DOCTYPE html PUBLIC "-//W3C//DTD HTML 4.01 Transitional//EN" "http://www.w3.org/TR/html4/loose.dtd">
<html>
<head>
<meta http-equiv="Content-Type" content="text/html;charset=UTF-8">
<title>Insert title here</title>
</head>
<body bgcolor="f0fff0">
<%
request.setCharacterEncoding("utf-8");
String username=request.getParameter("username");

out.println(username+"的爱好是:<br/>");
String[] hobby=request.getParameterValues("hobby");
for(String s:hobby)
{
 out.println(s+"</br>");
}
%>
</body>
</html>
```

图 3-6　显示用户选择的爱好的运行结果

4. request 对象的常用方法

① getProtocol()：获取请求使用的协议，如 HTTP/1.1。
② getServletPath()：获取请求的页面所在的位置。
③ getContentLength()：获取 HTTP 请求的长度。
④ getMethod()：获取表单提交信息的方式，如 POST 方法或 GET 方法。
⑤ getHeader(String s)：获取请求中的头的值。
⑥ getHeaderNames()：获取头名称的一个枚举。
⑦ getHeaders(String s)：获取头的全部值的一个枚举。
⑧ getRemoteAddr()：获取客户端的 IP 地址。
⑨ getRemoteHost()：获取客户端的名称。
⑩ getServerName()：获取服务器的名称。
⑪ getServerPort()：获取服务器的端口号。

【例 3-5】应用 request 对象的几个常用方法获取相关信息，参考代码如下，运行结果如图 3-7 所示。

教学视频

```
<%@ page language="java" contentType="text/html;charset=UTF-8" pageEncoding=
```

```jsp
"UTF-8"%>
    <!DOCTYPE html PUBLIC "-//W3C//DTD HTML 4.01 Transitional//EN" "http://www.w3.org/TR/html4/loose.dtd">
    <html>
    <head>
    <meta http-equiv="Content-Type" content="text/html;charset=UTF-8">
    <title>Insert title here</title>
    </head>
    <body bgcolor="f0fff0">
    <form action="" method=post name=form>
     <input type="submit" value="提交" name=submit>
    </form>
    <%
    String protocol=request.getProtocol();
    String path=request.getServletPath();
    String method=request.getMethod();
    String header=request.getHeader("accept");
    %>
    <table border=1>
    <tr>
    <td>客户使用的协议是:</td>
    <td>"<%=protocol %>"</td>
    </tr>

    <tr>
    <td>用户请求的页面所在的位置:</td>
    <td>"<%=path %>"</td>
    </tr>
    <tr>
    <td>客户提交信息的方式:</td>
    <td>"<%=method %>"</td>
    </tr>
    <tr>
    <td>获取http头文件中accept的值</td>
    <td>"<%=header %>"</td>
    </tr>
    </table>
    </body>
    </html>
```

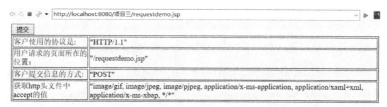

图 3-7 应用 request 对象的常用方法的运行结果

任务实施

教学视频

任务要求：创建两个页面，一个是考试试题页面 exam.jsp，另一个是考试结果页面 requestexam.jsp，用于展示考生的班级、姓名、成绩。下面分两步完成。

第一步：创建考试试题页面 exam.jsp。

```
<%@ page language="java" contentType="text/html;charset=UTF-8"
    pageEncoding="UTF-8"%>
<!DOCTYPE html PUBLIC "-//W3C//DTD HTML 4.01 Transitional//EN" "http://www.w3.org/TR/html4/loose.dtd">
<html>
<head>
<meta http-equiv="Content-Type" content="text/html;charset=UTF-8">
<title>Insert title here</title>
</head>
<body bgcolor="f0fff0">
<h1>Java Web 程序设计网上考试系统</h1>
<form action="requestexam.jsp" method="post">
请输入姓名：
<input type="text" name="name">
请选择班级：
<select name="class" size=1>
<option selected value="21 大数据 0031">21 大数据 0031
<option value="21 大数据 00312">21 大数据 0032
<option value="21 大数据 0033">21 大数据 0033
<option value="21 大数据 0034">21 大数据 0034
</select>
<br/><p>在 Java Web 中,可以获得用户表单提交的信息的内置对象是(    )<br/>
<input type="radio" name="t1" value="a"> response 对象
<input type="radio" name="t1" value="b">request 对象 <br/>
<input type="radio" name="t1" value="c"> session 对象
<input type="radio" name="t1" value="d">application 对象 <br/>
<p>Tomcat 的默认端口号是什么?(    )</p>
<input type="radio" name="t2" value="a"> 8080
<input type="radio" name="t2" value="b">80 <br/>
<input type="radio" name="t2" value="c">8009
<input type="radio" name="t2" value="d">8005 <br/>
<p>下面选项中,表示服务器错误的状态代码是(    )</p>
<input type="radio" name="t3" value="a">404
<input type="radio" name="t3" value="b">500 <br/>
<input type="radio" name="t3" value="c">302
<input type="radio" name="t3" value="d">100<br/>
<input type="submit" value="提交答案" name="submit">
</form>
</body>
</html>
```

第二步：创建考试结果页面 requestexam.jsp。

```jsp
<%@ page language="java" contentType="text/html;charset=UTF-8"
    pageEncoding="UTF-8"%>
<!DOCTYPE html PUBLIC "-//W3C//DTD HTML 4.01 Transitional//EN" "http://www.w3.org/TR/html4/loose.dtd">
<html>
<head>
<meta http-equiv="Content-Type" content="text/html;charset=UTF-8">
<title>Insert title here</title>
</head>
<body bgcolor="f0fff0">
<%
request.setCharacterEncoding("utf-8");
int n=0;
String strName=request.getParameter("name");
String strClass=request.getParameter("class");
String strTemp=strClass+"的"+strName+"Java Web 的考试成绩是:";
String s1=request.getParameter("t1");
String s2=request.getParameter("t2");
String s3=request.getParameter("t3");
if(s1==null)
{
    s1="";
}
if(s2==null)
{
    s2="";
}
if(s3==null)
{
    s3="";
}
if(s1.equals("b"))
{
    n=n+30;
}
if(s2.equals("a"))
{
    n=n+30;
}
if(s3.equals("b"))
{
    n=n+40;
}
%>
<%=strTemp %>
<p>您的得分为:<%=n %>分
```

```
</body>
</html>
```

任务拓展

1. 表单提交的 method 方法

表单提交的 method 方法分为两种。

（1）GET 方法

GET 方法的用法是将经过编码的表单内容通过 URL 发送，可以在地址栏看到传递的参数。

request1.jsp 页面的关键代码如下。

```
<body>
   <form action="qh.jsp" method="get">
   <input type="text" name="str">
   <input type="submit" value="提交">
   </form>
 </body>
```

qh.jsp 页面的关键代码如下。

```
<body>
<%
 request.setCharacterEncoding("utf-8");
 String s=request.getParameter("str");
out.print(s);
 %>
</body>
```

运行 qh.jsp，得到结果如图 3-8 所示。

图 3-8 可以在地址栏看到传递的参数

在图 3-8 的地址栏里可以看到 JSP 页面传递过来的 "hello"。

（2）POST 方法

POST 方法的用法是将表单内容通过 HTTP 发送，在地址栏看不到表单的提交信息，参考代码如下，运行结果如图 3-9 所示。

```
<body>
   <form action="qh.jsp" method="post">
   <input type="text" name="str">
   <input type="submit" value="提交">
   </form>
```

```
</body>
```

图 3-9 在地址栏看不到表单的提交信息

从图 3-9 可以看出，在把提交方法改为 POST 方法后，在地址栏看不到表单的提交信息。

2. 解决中文乱码问题

当应用 request.getParameter 得到 form 中的元素时，默认的字符编码方式为 ISO-8859-1，这种编码方式会使得 request 对象在获取用户提交的中文字符时出现乱码问题，所以必须对包含中文字符的信息采取特殊的处理方式，也就是转换字符编码，或使用 page 指令指定编码方式为 GB2312 或 UTF-8。下面介绍两种处理中文乱码的方式。

（1）转换字符编码

```
String str=request.getParameter("message");
Byte b[]=str.getBytes("ISO-8859-1");
str=new String(b);
```

（2）指定编码方式

在页面上使用 page 指令指定编码方式为 GB2312 或 UTF-8。request 对象在获取信息之前使用 setCharacterEncoding 方法设置编码方式的代码如下。

```
request.setCharacterEncoding("utf-8")
```

任务 2　应用 response 对象设计化妆品网站登录界面

 任务演示

用户先通过表单控件输入并提交信息，然后 JSP 获得表单数据进行逻辑处理，最后 JSP 根据处理结果转向不同的结果页面，登录界面、登录成功界面、登录失败界面分别如图 3-10、图 3-11、图 3-12 所示。

教学视频

图 3-10 登录界面

图 3-11 登录成功界面

图 3-12 登录失败界面

知识准备

1. response 对象重定向

response 对象可用于响应客户端请求，并向客户端输出信息。response 对象是 javax.servlet.http.HttpServletResponse 类的对象，它封装了 JSP 产生的响应，并将其发送到客户端以响应客户端的请求。请求的数据可以是各种类型的，甚至可以是文件。response 对象的常用方法如表 3-3 所示。

表 3-3 response 对象的常用方法

方法名	方法功能
addHeader (String name,String value)	添加 HTTP 文件头，该文件头将会传递到客户端
setHeader(String name,String value)	设置指定名称的 HTTP 文件头的值
addCookie (Cookie cookie)	添加一个 Cookie 对象，用来保存客户端的用户信息
encodeURL()	使用 sessionId 来封装 URL
flushBuffer()	强制将当前缓冲区的内容发送到客户端
sendRedirect(String location)	把响应发送到另一个指定的位置进行处理
getOutputStream()	返回客户端的输出流对象
setContentType()	动态改变 contentType 属性

（1）response 重定向

在 JSP 页面中，可以使用 response 对象中的 sendRedirect()方法将客户端请求重定向到一个不同的页面。

（2）<select></select>标记<option>标记

下拉式列表和滚动列表通过<select></select>标记和<option>标记来定义，<select></select>标记将<option>标记作为子标记，形成下拉列表或滚动列表的基本格式如下。

```
<select name="myName">
 <option value="item1">
 <option value="item2">
…
</select>
```

其中，在<select></select>标记中增加 size 属性的值就会变成滚动列表，size 属性的值是滚动列表的可见行的数目，滚动列表的基本格式如下。

```
<select name="myName"  size="正整数">
 <option value="item1">
 <option value="item2">
…
</select>
```

其中，request 对象通过参数 name 获取滚动列表中被选中的<option>标记的值（即参数 value 指定的值）。

【例 3-6】通过单击下拉列表的选项框实现重定向到对应的页面，参考代码如下，运行结果如图 3-13 所示。

教学视频

第一步：创建 response1.jsp 页面。

```
<%@ page language="java" contentType="text/html;charset=UTF-8"
    pageEncoding="UTF-8"%>
<!DOCTYPE html PUBLIC "-//W3C//DTD HTML 4.01 Transitional//EN" "http://www.w3.org/TR/html4/loose.dtd">
<html>
<head>
<meta http-equiv="Content-Type" content="text/html;charset=UTF-8">
<title>Insert title here</title>
</head>
<body>
    <form action="" method="post">
    <select name="opt">
    <option value="sina">新浪</option>
    <option value="sohu">搜狐</option>
    <option value="163">网易</option>
    </select>
    <input type="submit" value="go">
    </form>
</body>
</html>
```

第二步：创建 responseshow.jsp 页面。

```
<%@ page language="java" contentType="text/html;charset=UTF-8"
    pageEncoding="UTF-8"%>
<!DOCTYPE html PUBLIC "-//W3C//DTD HTML 4.01 Transitional//EN" "http://www.w3.org/TR/html4/loose.dtd">
<html>
<head>
<meta http-equiv="Content-Type" content="text/html;charset=UTF-8">
<title>Insert title here</title>
</head>
<body bgcolor="f0fff0">
 <% String str=request.getParameter("opt");
    if(str.equals("sina"))
    {response.sendRedirect("http://www.sina.com.cn");
    }
    else if(str.equals("sohu"))
    {response.sendRedirect("http://www.sohu.com");
    }
    else
    {response.sendRedirect("http://www.163.com");
```

```
        }
    %>
</body>
</html>
```

图 3-13 重定向到对应的页面的运行结果

2. response 对象刷新页面

【例 3-7】通过刷新页面来动态显示时间，每隔 2 秒刷新一次页面，参考代码如下，运行结果如图 3-14 所示。

```
<%@ page language="java" contentType="text/html;charset=UTF-8"
    pageEncoding="UTF-8"%>
<%@ page import="java.util.*" %>
<!DOCTYPE html PUBLIC "-//W3C//DTD HTML 4.01 Transitional//EN" "http://www.w3.org/TR/html4/loose.dtd">
<html>
<head>
<meta http-equiv="Content-Type" content="text/html;charset=UTF-8">
<title>Insert title here</title>
</head>
<body>
<body bgcolor="f0fff0">
<p>当前时间是:<br>
<%
out.print(""+new Date().toLocaleString());
response.setHeader("Refresh","2");
%>
</body>
</html>
```

当前时间是:
2022-6-16 14:57:06

图 3-14 动态显示时间的运行结果

任务实施

第一步：编写 head.jsp 页面，并将 href 的属性值设置为 "login.jsp"。

```
<%@ page language="java" contentType="text/html;charset=UTF-8" pageEncoding="UTF-8"%>
```

```html
<!DOCTYPE html PUBLIC "-//W3C//DTD HTML 4.01 Transitional//EN" "http://www.w3.org/TR/html4/loose.dtd">
<html>
<head>
<meta http-equiv="Content-Type" content="text/html;charset=UTF-8">
<title>Insert title here</title>
</head>
<style type="text/css">
#bd{background:url("image/back1.jpg");
            background-size:100%;
}
</style>
<body id=bd>
<div align="center">
  <table cellSpacing="1" cellPadding="1" width="660" align="center" border="0">
   <tr valign="bottom">
      <td><a href="login.jsp">登录</a></td>
      <td><a href="">注册</a></td>
      <td><a href="">浏览化妆品</a></td>
      <td><a href="">查找化妆品</a></td>
      <td><a href="">查看购物车</a></td>
       <td><a href="">查看订单</a></td>
       <td><a href="">退出</a></td>
       <td><a href="">主页</a></td>
    </tr>
   </table>
  </div>
</body>
</html>
```

第二步：编写 index.jsp 页面。

```jsp
<%@ page language="java" contentType="text/html;charset=UTF-8" pageEncoding="UTF-8"%>

<!DOCTYPE html PUBLIC "-//W3C//DTD HTML 4.01 Transitional//EN" "http://www.w3.org/TR/html4/loose.dtd">
<html>
<head>
<meta http-equiv="Content-Type" content="text/html;charset=UTF-8">
<title>Insert title here</title>
</head>
<%@ include file="head.jsp" %>
<body bgcolor="f0fff0">
<img alt="" src="image/a1.JPG">
</body>
</html>
```

第三步：编写 login.jsp 页面。

```jsp
<%@ page language="java" contentType="text/html;charset=UTF-8" pageEncoding="UTF-8"%>

<!DOCTYPE html PUBLIC "-//W3C//DTD HTML 4.01 Transitional//EN" "http://www.w3.org/TR/html4/loose.dtd">
<html>
<head>
<meta http-equiv="Content-Type" content="text/html;charset=UTF-8">
<title>Insert title here</title>
</head>
<%@ include file="head.jsp" %>
<body background="image/back1.jpg">
<center>
<form action="login2.jsp" method="post">
<br><br>
<table border="2">
<tr>
   <th colspan=2>请您登录</th>
</tr>
<tr>
<td>登录名称</td>
   <td><input type="text" name="logname"></td>
</tr>
<tr>
   <td>输入密码</td>
   <td><input type="password" name="password"></td>
</tr>
</table>
<td><input type="submit" value="提交"></td>
</form>
</center>
</body>
</html>
```

第四步：修改 login2.jsp 页面。

```jsp
<%@ page language="java" contentType="text/html;charset=UTF-8" pageEncoding="UTF-8"%>

<!DOCTYPE html PUBLIC "-//W3C//DTD HTML 4.01 Transitional//EN" "http://www.w3.org/TR/html4/loose.dtd">
<html>
<head>
<meta http-equiv="Content-Type" content="text/html;charset=UTF-8">
<title>Insert title here</title>
</head>
```

```jsp
<body bgcolor="f0fff0">
<%request.setCharacterEncoding("utf-8");
String username=request.getParameter("logname");
String pwd=request.getParameter("password");
if (username.equals("liaoli")&&pwd.equals("123"))
response.sendRedirect("success.jsp");
else
response.sendRedirect("fail.jsp");
%>
</html>
```

第五步:新建一个 success.jsp 页面,当登录成功时,success.jsp 页面会输出"登录成功,欢迎你进入化妆品网站!"

```jsp
<%@ page language="java" contentType="text/html;charset=UTF-8" pageEncoding="UTF-8"%>
<!DOCTYPE html PUBLIC "-//W3C//DTD HTML 4.01 Transitional//EN" "http://www.w3.org/TR/html4/loose.dtd">
<html>
<head>
<meta http-equiv="Content-Type" content="text/html;charset=UTF-8">
<title>Insert title here</title>
</head>
<%@ include file="head.jsp" %>
<body bgcolor="f0fff0">
<b>登录成功,欢迎你进入化妆品网站!</b>
</body>
</html>
```

第四步:新建一个 fail.jsp 页面,当登录失败时,fail.jsp 页面会输出"你输入的用户名或密码有误!"

```jsp
<%@ page language="java" contentType="text/html;charset=UTF-8" pageEncoding="UTF-8"%>
<!DOCTYPE html PUBLIC "-//W3C//DTD HTML 4.01 Transitional//EN" "http://www.w3.org/TR/html4/loose.dtd">
<html>
<head>
<meta http-equiv="Content-Type" content="text/html;charset=UTF-8">
<title>Insert title here</title>
</head>
<%@ include file="head.jsp" %>
<body bgcolor="f0fff0">
<b>你输入的用户名或密码有误!</b>
</body>
</html>
```

任务拓展

当服务器对客户端请求做出响应时,它发送的首行内容被称为状态行。状态行包括由 3 位数字组成的状态代码和状态代码的描述,常用的状态代码如表 3-4 所示。

表 3-4 常用的状态代码

状态代码	状态代码的描述
101	服务器正在升级协议
100	客户端可以继续提出请求
201	请求成功且在服务器上创建了新的资源
202	请求已被接收,但还没有处理完毕
200	请求成功
203	客户端给出的元信息不是发自服务器的
204	请求成功,但没有新信息
205	客户端必须重置文档视图
206	服务器执行了部分 GET 请求
300	请求的资源有多种表示方法
301	资源已经被永久移动到新位置
302	资源已经被临时移动到新位置
303	响应可以在另外一个 URL 中找到
304	GET 请求不可用
305	请求必须通过代理来访问
400	请求有语法错误
401	请求需要 HTTP 认证
403	服务器拒绝请求
404	请求的资源不可用
405	请求的方法是不允许使用的
406	请求的资源只能用请求来响应,但不能用请求的内容特性来响应
407	客户端必须得到认证
408	请求超时
409	发生冲突,请求不能完成
410	请求的资源已经不可用
411	请求需要定义一个内容长度才能被处理
413	请求数量太大,请求被拒绝
414	请求的 URL 太大
415	请求的格式被拒绝

在出现问题时,一般不需要修改状态行,服务器会自动响应,并发送相应的状态代码。我们也可以使用 response 对象的 setStatus(int n)方法来改变响应的状态行的内容。

任务 3　应用 session 对象设计火锅点餐系统

任务演示

"日暮长街吃火锅，家家扶得醉人归"，这句诗恰如其分地描述了火锅的受欢迎程度。下面我们来设计一个火锅点餐系统，其界面如图 3-15 所示。

a. 火锅点餐系统首页

b. 火锅点餐系统点餐页面　　　　　　c. 火锅点餐系统结账页面

图 3-15　火锅点餐系统界面

知识准备

HTTP 协议是一种无状态协议。用户先向服务器发出请求（request 对象），然后服务器返回响应（response 对象），服务器不保留连接的有关信息，因此在下一次建立连接时，服务器已没有以前的连接信息了，无法判断这一次建立的连接和以前建立的连接是否属于同一用户。而 session 可以将会话数据保存在服务器内，本节将对 session 进行详细讲解。

1. session 对象

session 对象是与请求相关的 HttpSession 对象，它封装了属于客户端会话的所有信息，session 对象也是一个 JSP 内置对象，它会在第一个 JSP 页面被装载时自动创建，完成会话期管理。当用户在应用程序的 Web 页面之间跳转时，存储在 session 对象中的变量不会丢失，在整个用户会话中都一直存在。当用户发出请求时，如果该用户没有会话，服务器将自动创建一个 session 对象，当会话过期或被放弃后，服务器将终止该会话。

2. session 对象的 id

session 对象被分配了一个字符串类型的 id，服务器会将这个 id 发送到客户端，并存放在用户的 Cookie 中。这样，session 对象和用户之间就建立起一一对应的关系，即每个用户都对应一个 session 对象（也称用户的会话），不同用户的 session 对象互不相同，且具有不同的 id。简单地说，用户在访问一个 Web 服务目录期间，服务器会为该用户分配一个 session 对象，并且服务器可以在各个页面使用这个 session 对象记录当前用户的有关信息，服务器能保证不同用户的 session 对象互不相同。

3. session 对象的常用方法

session 对象常用下列方法处理数据。

① public void setAttribute (String name, Object value)：该方法将参数 Object 指定的对象 value 添加到 session 对象中，并为添加的对象指定了一个索引关键字。其中 name 指定 session 属性的名称，value 绑定 session 属性的 name 值，该值是一个对象。

② public Object getAttribute(String name)：获取 session 作用域里指定属性的值，该值是 Object 类型的。

③ public void removeAttribute(String name)：指定会话中需要移除的属性名称。

④ public Enumeration getAttributeNames()：返回会话中所有 session 属性的枚举。

⑤ public void invalidate()：销毁当前的 session 对象。

⑥ public String getId()：返回当前 session 的 id。

⑦ public long getCreationTime()：返回当前会话的创建时间，单位是秒。

⑧ public long getLastAccessedTime()：返回当前会话的最后一次访问时间。

⑨ public void setMaxInactiveInterval(int interval)：设置 session 的有效时间，单位是秒。

⑩ public void getMaxInactiveInterval()：返回 session 的失效时间，如果返回值为-1，则表示永不过期，单位是秒。

教学视频

【例 3-8】获取 session 对象的 id，参考代码如下，运行结果如图 3-16 所示。

```
<body bgcolor="f0fff0">
我是第一个页面
<%String id=session.getId();
out.print("第一个页面的session的id是:<br>"+id);
%>
<form action=" " method="post">
<input type="text" name="my">
 <input type="submit" value="送出" name="submit">
</form>
```

图 3-16 获取 session 对象的 id

【例 3-9】打开家电购物商城网站，如果当前没有完成登录，则跳转到登录窗口；如果当前已经完成登录，则显示当前用户的信息；如果从登录窗口进入主页，则设置当前用户会话的用户名，参考代码如下，家电购物商城网站的登录窗口和主页分别如图 3-17、图 3-18 所示。

参考代码如下：

（1）session2.jsp

```html
<body>
<form action="session21.jsp" method="post">
<table>
<tr>
<td>用户名:</td><td><input type="text" name="user"/> </td>
</tr>
<tr>
<td>密码:</td><td><input type="password" name="pass"/></td> </tr>
<tr ><td align="right"><input type="submit" value="提交"/> </td>
<td align="center"><input type="reset" value="重置"/></td></tr>
</table>
</form>
</body>
```

（2）session21.jsp

```jsp
<body>
<% request.setCharacterEncoding("utf-8");
String user=request.getParameter("user");
if(!(user==""||user.equals("")))
{
    session.setAttribute("user", user);
}
%>
<%
user=(String)session.getAttribute("user");
if("".equals(user)||user==null)
    {out.println("用户未登录,即将跳转到登录界面");
response.setHeader("Refresh","5");
response.sendRedirect("session2.jsp");
}
else
{ out.println("家电购物商城</br></br>");
    out.println("欢迎"+user+"光临本商场</br>?");
}
%>
</body>
```

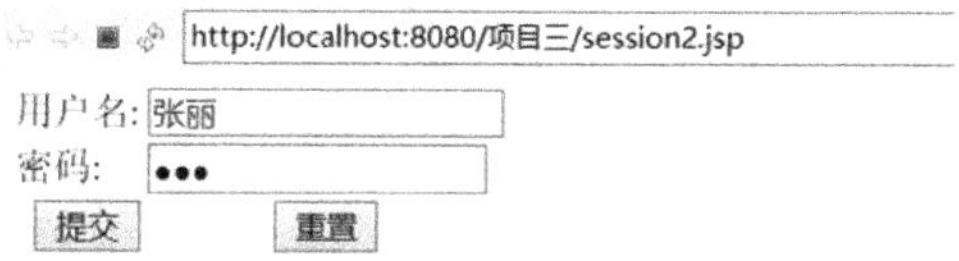

图 3-17　登录窗口 　　　　　　　　图 3-18　主页

【例 3-10】应用 session 对象的方法获取 session 对象的 id、创建时间、最大有效时间、最后操作时间，代码如下，运行结果如图 3-19 所示。

教学视频

```
<body>
<h1>会话信息</h1>
<table border="1">
<tr>
<th>session 属性</th>
<th>值</th>
</tr>
<tr>
<td>session 的 id</td>
<td><%=session.getId() %></td>
</tr>
<tr>
<td>session 的创建时间</td>
<td><%=session.getCreationTime() %></td>
</tr>
<tr>
<td>session 的最大有效时间</td>
<td><%=session.getMaxInactiveInterval() %></td>
</tr>
<tr>
<td>session 的最后操作时间</td>
<td><%=session.getLastAccessedTime() %></td>
</tr>
</table>
</body>
```

http://localhost:8080/项目三/session3.jsp

会话信息

session 属性	值
session 的 id	69A57BA6AA6B7286FC804D4FFE72E976
session 的创建时间	1653914365064
session 的最大有效时间	1800
session 的最后操作时间	1653914365064

图 3-19 应用 session 对象的方法的运行结果

（注：此处显示的时间为任意输入的数值，在此仅供示范，用户可根据实际情况输入真实的时间）

任务实施

教学视频

任务要求：用 session 对象实现点餐、存储用户名和显示点餐信息的功能，并设计一个结账功能。

第一步：创建一个 buy1.jsp 页面，用于输入用户信息。

```
<head>
<br> 输入姓名:<a href="username.jsp">输入用户名</a>
```

```
<br>点餐:<a href="order.jsp">点餐</a>
<br>结账:<a href="bill.jsp">结账</a>
</head>
<!--创建了一个表单,输入用户名和一个提交按钮-->
<body bgcolor="f0fff0">    <br><br>输入用户名
 <form action="" method="post" name="form">
   <input type="text" name="name">
   <input type="submit" value="确定" name=submit>
 </form>

<!--请求获取输入的用户名,并赋给 name-->
<% String name=request.getParameter("name");
 if(name==null)
 name="";
 else
//将用户名保存到 session 对象的 name 中
 session.setAttribute("name",name);%>
</body>
```

第二步:创建一个 order.jsp 页面用于点餐。

```
<head>
<br> 输入姓名:<a href="username.jsp">输入用户名</a>
<br>点餐:<a href="order.jsp">点餐</a>
<br>结账:<a href="bill.jsp">结账</a>
</head>
<body bgcolor="f0fff0">
<br><br>请点餐:
<input type ="checkbox" name="data" value="猪肉 35.4 元">猪肉 35.4 元<br> <!--输入框
的类型是 checkbox,name 表示复选框的名称,value 表示复选框的值-->
<input type ="checkbox" name="data" value="羊肉 40.2 元">羊肉 40.2 元<br>
<input type ="checkbox" name="data" value="牛肉 48.3 元">牛肉 48.3 元<br>
<input type ="checkbox" name="data" value="山药 20 元">山药 20 元<br>
<input type ="checkbox" name="data" value="小菜 15.5 元">小菜 15.5 元<br><input
type="submit" value="提交">
<input type="reset" value="重设">
</form>
<%String order[]=request.getParameterValues("data"); //请求获取复选框 data 的值,也就
是点餐的信息
   if(order!=null)
   {
StringBuffer str=new StringBuffer(); //创建一个字符串变量
for(int k=0;k<order.length;k++)
{
str.append(order[k]+"<br>"); //将点餐信息放到 str 字符串变量中
}
session.setAttribute("order",str); //将 str 的值存储到 session 对象的 order 关键字中
}
```

```
%>
</body>
```

第三步:创建一个 bill.jsp 页面用于结账。

```
bill.jsp
<br> 输入姓名:<a href="username.jsp">输入用户名</a>
<br>点餐:<a href="order.jsp">点餐</a>
<br>结账:<a href="bill.jsp">结账</a>
</head>
<%! public String handleStr(String s)
{
try{
byte[] bb=s.getBytes("iso-8859-1");
s=new String(bb);
}catch(Exception e)
{   }
return s;
}
%>
<body bgcolor="f0fff0">
<% String username=(String)session.getAttribute("name");
if(username==null||username.length()==0)
{   out.print("返回到输入姓名");} %>
<%StringBuffer order=(StringBuffer)session.getAttribute("order");
String order1=new String(order);
double sum=0;
String[] price=order1.split("[^0123456789.]");
if(price!=null)
{   for(String item:price)
{   try{  sum=sum+Double.parseDouble(item);
}catch(NumberFormatException e2)
{ } } }
%>  <br><br><%=handleStr(username) %>点的餐:<br>
<%=handleStr(order1) %><br>
总付款:<%=sum %>
```

任务拓展

session 对象失效

　　有两种方法可以让 session 对象失效,分别是通过"超时限制"使 session 对象失效和强制让 session 对象失效。Web 服务器采用"超时限制"判断客户端是否还在继续访问。在一定时间内,如果某个客户端一直没有请求访问,那么 Web 服务器就会认为该客户端已经结束请求,并且将与该客户端会话所对应的 HttpSession 对象变成垃圾对象,等待垃圾收集器将其从内存中彻底清除。通过"超时限制"管理会话生命周期的方法如表 3-5 所示。此外,还可以使用 invalidate()方法强制让 session 对象失效。

表 3-5　通过"超时限制"管理会话生命周期的方法

getLastAccessedTime()	返回客户端最后一次发送与这个会话相关联的请求的时间
getMaxInactiveInterval()	以秒为单位返回一个会话内的两个请求的最大时间间隔，Servlet 容器会在用户访问期间保存这个会话，并处于打开状态
setMaxInactiveInterval (int interval)	以秒为单位，指定在无任何操作的情况下，Session 失效的时间，也就是超时时间

任务 4　应用 application 对象设计留言板

  任务演示

教学视频

我们一直重视粮食安全，并提倡"厉行节约、反对浪费"的社会风尚，现在通过留言板来收集有关节约食粮的主题信息，留言板界面、留言信息提交界面、留言信息展示界面分别如图 3-20、图 3-21、图 3-22 所示。

图 3-20　留言板界面

图 3-21　留言信息提交界面　　　　图 3-22　留言信息展示界面

  知识准备

1. application 对象

application 对象可用来保存所有应用程序中的公有数据。在服务器启动并且自动创建 application 对象后，只要服务器没有关闭，或是预设时间没有超时，则 application 对象一直存在，且所有用户都可以共享 application 对象。application 对象是一个实现了 ServletContext 接口的类对象，它提供了一些方法与 Web 服务器进行信息传递。

application 对象的常用方法有如下几种。

① public void setAttribute(String key,Object obj)：以键值对的方式将一个对象的值存放到 application 对象中。

② public Object getAttribute(String key)：根据名称获取 application 对象中存放的对象的值，返回值的数据类型是 Object 类型。在实际应用时要视其数据类型进行转换。

③ public void removeAttribute(String name)：指定要移除的属性的名称。

④ public Enumeration getAttributeNames()：用于获取应用程序作用域范围内的所有属性的枚举。

2. Vector 类

Vector 类是一元集合，可以加入重复数据，它的作用和数组类似，可以保存一系列数据。它的优点是可以很方便地对集合内的数据进行查找、增加。Vector 类的方法可分为以下两类。

（1）Vector 类的三个构造方法

```
public Vector(int initialCapacity,int capacityIncrement)
public Vector(int initialCapacity)
public Vector()
```

（2）Vector 类常用的方法

add(Object obj)：把组件添加到向量尾部，同时向量长度加 1。

addElementAt(Object obj，int index)：在 Vector 类的结尾处添加元素。

size()：返回 Vector 类的元素总数。

elementAt(int index)：取得特定位置的元素，返回值为整型数据。

setElementAt(Object obj，int index)：重新设定指定位置的元素。

removeElementAt(int index)：删除指定位置的元素。

教学视频

【例 3-11】创建一个空的 Vector 类，并实现添加元素、移除元素、输出元素的功能，参考代码如下，输出 Vector 类的所有元素的运行结果如图 3-23 所示。

```jsp
<%@ page language="java" contentType="text/html;charset=UTF-8"
    pageEncoding="UTF-8"%>
<%@ page import="java.util.*" %>
<!DOCTYPE html PUBLIC "-//W3C//DTD HTML 4.01 Transitional//EN" "http://www.w3.org/TR/html4/loose.dtd">
<html>
<head>
<meta http-equiv="Content-Type" content="text/html;charset=UTF-8">
<title>Insert title here</title>
</head>
<body>
<%
Vector v=new Vector(); //创建空的 Vector 类
for(int i=0;i<4;i++)
{
    v.add(new String("幸运星"+i));
}
v.remove(1); //移除索引值为 1 的元素
```

```
for(int i=0;i<v.size();i++)
{
    out.print("幸运星"+v.indexOf(v.elementAt(i))+":"+v.elementAt(i)+"    ");
}
%>
</body>
</html>
```

http://localhost:8080/项目三/VectorDemo.jsp

幸运星0:幸运星0 幸运星1:幸运星2 幸运星2:幸运星3

图 3-23　输出 Vector 类的所有元素的运行结果

3. <textarea></textarea>标记

<textarea></textarea>标记可用于定义多行的文本输入控件。

其常用的属性如下。

① cols：规定文本区内的可见列数。

② rows：规定文本区内的可见行数。

 任务实施

教学视频

任务要求：

① 用户通过 input.jsp 页面向 messagepane.jsp 页面输入名字、留言标题和留言。

② 在 messagepane.jsp 页面获取这些内容后，先用同步方法将这些内容添加到一个向量中，然后再将这个向量添加到 application 对象中。

③ 当用户单击"查看留言板"按钮时，show.jsp 页面负责显示所有用户的留言，从 application 对象中取出向量并遍历向量中存储的信息。

第一步：创建 input.jsp 页面，用于输入名字、留言标题和留言。

```
<body bgcolor="f0fff0">
<form action="messagepane.jsp" method="post" name="form">
输入名字:<br><input type="text" name="name"><br>
留言标题:<br><input type="text" name="title"><br>
留言:<br><textarea rows="10" cols="36" name="messages">
</textarea>
<br><input type="submit" value="提交信息" name="submit">
</form>
<form action="show.jsp" method="post" name="form1">
<input type="submit" value="查看留言板" name="look">
</form>
</body>
```

第二步：创建 messagepane.jsp 页面，用于提交留言的相关信息。

```
<body bgcolor="f0fff0">
<%! Vector v=new Vector();
    int i=0;
```

```
    ServletContext application;
    synchronized void leaveWord(String s)
    {   application=getServletContext();
    i++;   v.add("no."+i+","+s);   application.setAttribute("mess",v);   }%>
<% String name=request.getParameter("name");
String title=request.getParameter("title");
String messages=request.getParameter("messages");
if(name==null)
name="无名";
if(title==null)
title="无标题";
if(messages==null)
messages="无信息";
String s=name+"#"+title+"#"+messages;
leaveWord(s);
out.print("你的信息已经提交!");%>
<a href="input.jsp">返回留言页面</a>
</body>
```

第三步：创建 show.jsp 页面，用于显示所有的留言人、留言标题、留言内容。

```
<%! public String handlestr(String s)
{
try{
byte[] bb=s.getBytes("iso-8859-1");
s=new String(bb);
}catch(Exception e)
{}
return s;}
%>
<body bgcolor="f0fff0">
<%Vector v=(Vector)application.getAttribute("mess");
for(int i=0;i<v.size();i++)
{String message=(String)v.elementAt(i);
String[] a=message.split("#");
out.print("留言人:"+handlestr(a[0])+",");
    out.print("标题:"+handlestr(a[1])+"<br>");
    out.print("留言内容:<br>"+handlestr(a[2]));
    out.print("<br>---------------------<br>");}%></body>
```

任务拓展

getInitParameter 方法和 getInitParameterNames 方法的应用

getInitParameter 方法和 getInitParameterNames 方法可用于获取应用程序的参数，应用程序的参数被配置在 web.xml 文件中，被放在 "</web-app>" 前面，使用的节点是 "<context-param>"。

① public String getInitParameter(String name)：使用 getInitParameter 方法返回指定属性对应的应用程序的初始值，如果没有初始值就返回 null。

② public Enumeration getInitParameterNames()：使用 getInitParameterNames 方法获取所有初始化参数的枚举。

【例3-12】在应用程序中设置作者信息，并显示在页面上。

第一步：在 web.xml 文件中配置如下参数。

```
<?xml version="1.0" encoding="UTF-8"?>
<web-app xmlns:xsi="http://www.w3.org/2001/XMLSchema-instance" xmlns="http://xmlns.jcp.org/xml/ns/javaee" xsi:schemaLocation="http://xmlns.jcp.org/xml/ns/javaee http://xmlns.jcp.org/xml/ns/javaee/web-app_3_1.xsd" id="WebApp_ID" version="3.1">
    <context-param>
    <param-name>author</param-name>
    <param-value>丽丽</param-value>
    </context-param>
</web-app>
```

第二步：编写 getInitParameterdemo.jsp 页面的代码。

```
<%@ page language="java" contentType="text/html;charset=UTF-8"
    pageEncoding="UTF-8"%>
<!DOCTYPE html PUBLIC "-//W3C//DTD HTML 4.01 Transitional//EN" "http://www.w3.org/TR/html4/loose.dtd">
<html>
<head>
<meta http-equiv="Content-Type" content="text/html;charset=UTF-8">
<title>Java Web 项目开发</title>
</head>
<body>
本书的作者是:<%=application.getInitParameter("author") %>
</body>
</html>
```

运行结果如图 3-24 所示。

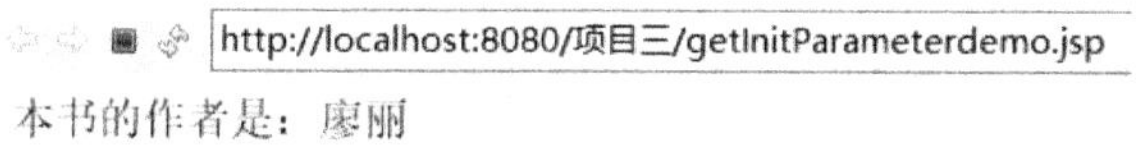

http://localhost:8080/项目三/getInitParameterdemo.jsp

本书的作者是：廖丽

图 3-24　设置作者信息并显示在页面上的运行结果

【例3-13】《神童诗·四喜》中写到人生四大喜事："久旱逢甘雨，他乡遇故知。洞房花烛夜，金榜题名时"。下面编写程序，在应用程序中初始化上述的人生四大喜事，并显示在页面上，显示效果如图 3-25 所示。

第一步：在 web.xml 文件中配置如下参数。

```
<?xml version="1.0" encoding="UTF-8"?>
<web-app xmlns:xsi="http://www.w3.org/2001/XMLSchema-instance" xmlns="http://xmlns.jcp.org/xml/ns/javaee" xsi:schemaLocation="http://xmlns.jcp.org/xml/ns/javaee http://xmlns.jcp.org/xml/ns/javaee/web-app_3_1.xsd" id="WebApp_ID" version="3.1">
```

```xml
<context-param>
<param-name>第一喜</param-name>
<param-value>久旱逢甘雨</param-value>
</context-param>
<context-param>
<param-name>第二喜</param-name>
<param-value>他乡遇故知</param-value>
</context-param>
<context-param>
<param-name>第三喜</param-name>
<param-value>洞房花烛夜</param-value>
</context-param>
<context-param>
<param-name>第四喜</param-name>
<param-value>金榜题名时</param-value>
</context-param>
</web-app>
```

第二步：编写 getInitParameterNamesdemo.jsp 页面的代码。

```jsp
<%@page import="java.util.Enumeration"%>
<%@ page language="java" contentType="text/html;charset=UTF-8"
    pageEncoding="UTF-8"%>
<!DOCTYPE html PUBLIC "-//W3C//DTD HTML 4.01 Transitional//EN" "http://www.w3.org/TR/html4/loose.dtd">
<html>
<head>
<meta http-equiv="Content-Type" content="text/html;charset=UTF-8">
<title>Insert title here</title>
</head>
<body>
<% Enumeration<String> paraNames=application.getInitParameterNames();
while(paraNames.hasMoreElements())
{
    String name=paraNames.nextElement();
    String value=application.getInitParameter(name);
    out.print(name+":"+value+"</br>");
}
%>
</body>
</html>
```

```
http://localhost:8080/项目三/getInitParameterNamesdemo.jsp
第一喜:久旱逢甘雨
第二喜:他乡遇故知
第三喜:洞房花烛夜
第四喜:金榜题名时
```

图 3-25　人生四大喜事的显示效果

任务 5 应用 Cookie 对象制作站点计数器

任务演示

应用 Cookie 对象制作站点计数器,当用户第一次访问某一网页时,id 可通过 Cookie 对象传送给用户,当用户再次访问该网页时,该 id 对应的计数器的数值增加 1,最终统计出用户的访问次数,站点计数器的运行结果如图 3-26 所示。

教学视频

图 3-26 站点计数器的运行结果

知识准备

Cookie 是服务器在 HTTP 响应中附带传给浏览器的一个文本文件,一旦浏览器保存了某个 Cookie,那么在之后的请求和响应过程中,就会将此 Cookie 来回传递,这样就可以通过 Cookie 这个载体完成客户端和服务器的数据交互。

Cookie 给网站和用户带来的好处非常多,具体有如下几点。

① Cookie 能使站点跟踪特定访问者的访问次数、最后的访问时间和访问者进入站点的路径。

② Cookie 能告诉在线广告商其广告被点击的次数,从而更精确地投放广告。

③ 当未到 Cookie 有效期限时,用户无须重复输入密码和用户名,就可以直接进入曾经浏览过的一些站点。

④ Cookie 能帮助站点统计用户个人资料,实现个性化服务。

1. Cookie 对象的创建

可以调用 Cookie 对象的构造函数创建 Cookie 对象。Cookie 对象的构造函数有两个字符串参数:Cookie 对象的名称和 Cookie 对象的值。创建 Cookie 对象的语法格式如下。

```
Cookie cookie = new Cookie("name","tom");
response.addCookie(cookie);
```

2. Cookie 对象的读取

要读取保存到客户端的 Cookie 对象,可以使用 request 对象的 getCookies()方法将所有客户端传来的 Cookie 对象以数组形式排列。如果要取出符合需要的 Cookie 对象,就需要循环比较数组内每个对象的关键字,基本方法如下。

```
Cookie[] cookies = request.getCookies();
for (Cookie cookie:cookies){
 out.write(cookie.getName()+":"+cookie.getValue()+"<br/>");
}
```

3. Cookie 的常用方法

Cookie 的常用方法如表 3-6 所示。

表 3-6 Cookie 的常用方法

方法名	功能
void setMaxAge (int age)	设置 Cookie 的有效时间，单位为秒，无返回值
int getMaxAge()	获取 Cookie 的有效时间，有返回值
String getName()	获取 Cookie 的名称
String getValue()	获取 Cookie 的值

【例 3-14】Cookie 对象的创建与遍历。

第一步：创建 Cookiedemo.jsp 页面。

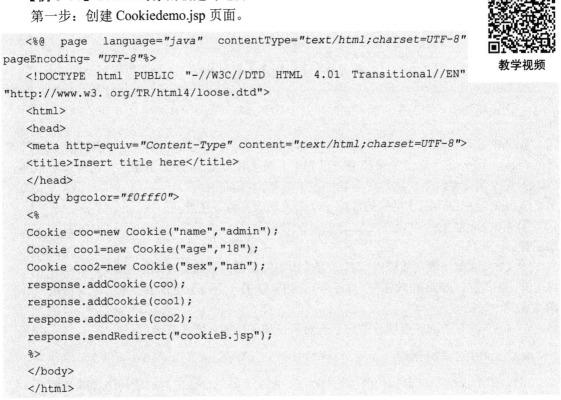

教学视频

```jsp
<%@ page language="java" contentType="text/html;charset=UTF-8" pageEncoding= "UTF-8"%>
<!DOCTYPE html PUBLIC "-//W3C//DTD HTML 4.01 Transitional//EN" "http://www.w3.org/TR/html4/loose.dtd">
<html>
<head>
<meta http-equiv="Content-Type" content="text/html;charset=UTF-8">
<title>Insert title here</title>
</head>
<body bgcolor="f0fff0">
<%
Cookie coo=new Cookie("name","admin");
Cookie coo1=new Cookie("age","18");
Cookie coo2=new Cookie("sex","nan");
response.addCookie(coo);
response.addCookie(coo1);
response.addCookie(coo2);
response.sendRedirect("cookieB.jsp");
%>
</body>
</html>
```

第二步：遍历 Cookie 对象。

```jsp
<%@ page language="java" contentType="text/html;charset=UTF-8" pageEncoding="UTF-8"%>
<!DOCTYPE html PUBLIC "-//W3C//DTD HTML 4.01 Transitional//EN" "http://www.w3.org/TR/html4/loose.dtd">
<html>
<head>
<meta http-equiv="Content-Type" content="text/html;charset=UTF-8">
<title>Insert title here</title>
</head>
<body bgcolor="f0fff0">
```

```
<%
Cookie[] coo=request.getCookies();
for(Cookie c:coo)
{
    out.println(c.getName()+"  "+c.getValue());
}
%>
</body>
    </html>
```

运行结果如图 3-27 所示。

http://localhost:8080/项目三/cookieB.jsp

name admin age 18 sex nan

图 3-27　Cookie 对象的创建与遍历的运行结果

 任务实施

应用 Cookie 对象制作站点计数器的代码如下，具体操作可扫描侧方二维码。

教学视频

```
<%@ page language="java" contentType="text/html;charset=UTF-8" pageEncoding="UTF-8"%>
<!DOCTYPE html PUBLIC "-//W3C//DTD HTML 4.01 Transitional//EN" "http://www.w3.org/TR/html4/loose.dtd">
<html>
<head>
<meta http-equiv="Content-Type" content="text/html;charset=UTF-8">
<title>用 Cookie 对象制作一个计数器</title>
</head>
<body>
<%
    Cookie thisCookie = null;
     boolean cookieFound = false;
    // 从请求中获取 Cookies
    Cookie[] cookies = request.getCookies();
    if(cookies!=null)
    {
     for(int i=0;i < cookies.length;i++)
     {
         thisCookie = cookies[i];
         //检查是否存在 CookieCount 数据
         if (thisCookie.getName().equals("CookieCount"))
         {
             cookieFound = true;
             break;
         }
     }
    }
    //输出页面
```

```
    out.println("<center><h1>站点计数器</h1></center>");
     // 显示客户端详细信息
    if (cookieFound)
    {
      // 获取 Cookie 的值,并对其做加 1 运算
      int cookieCount = Integer.parseInt(thisCookie.getValue());
      cookieCount++;
      out.println("<font color=blue size=+1>");
      out.println("<p>这是 1 分钟内第<B> " +cookieCount +"</B>次访问本页面\n");
      // 设置 Cookie 的新值,加到相应的对象中
      thisCookie.setValue(String.valueOf(cookieCount));
      thisCookie.setMaxAge(60*1);
      response.addCookie(thisCookie);
    }
    if (cookieFound == false)
    {
      out.println("<font color=blue size=+1>");
      out.println("<p>你在近 1 分种没有访问过此页面,现在是第 1 次访问此页面");
           // 创建新的 Cookie 并设置存活期
           thisCookie = new Cookie("CookieCount","1");
           thisCookie.setMaxAge(60*1);
           // 在 response 对象中加入 Cookie
           response.addCookie(thisCookie);
    }
  %>
  </body>
  </html>
```

任务拓展

Cookie 与 session 的区别

JSP 中的 Cookie 与 session 的区别具体体现在如下几点。

① Cookie 保存在客户端,所以可以长期保存,而 session 是不可以长期保存的。

② JSP 中的 session 保存在服务器中,客户端不知道其中的信息,而 Cookie 保存在客户端中,服务器可以知道其中的信息。

③ JSP 中的 session 保存的是对象,Cookie 保存的是字符串。

④ JSP 中的 session 不可以区分路径,在同一个用户访问同一个网站期间,所有的 session 在任何地方都可以被访问。但如果 Cookie 设置了路径参数,那么同一个网站、不同路径的 Cookie 是互相访问不到的。

项 目 实 训

实训一 根据家庭生活采购账单计算消费总额

要求:通过表单提交一份家庭生活采购账单,计算消费总额,结果如图 3-28 所示。

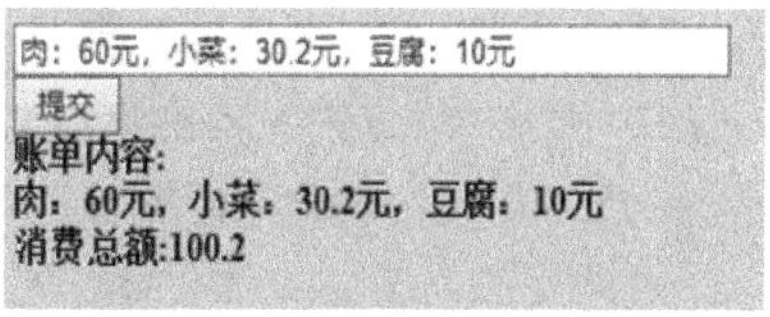

图 3-28 根据家庭生活采购账单计算消费总额的结果

实训二 设计调查问卷主界面

请应用所学知识完成调查问卷设计，要求输入学生的姓名和性别，显示输入的学生信息，并将调查结果展示出来，调查问卷主界面如图 3-29 所示。

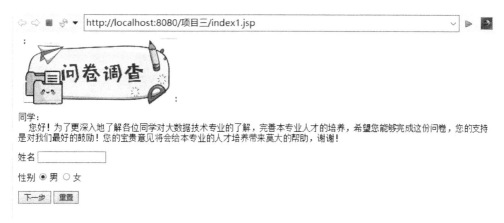

图 3-29 调查问卷主界面

实训三 设计用户注册程序

要求：设计用户注册程序 register.jsp 和 showreg.jsp，注册信息包括用户名、年龄、性别。将注册信息提交到 showreg.jsp 进行注册和检验，若用户名为"admin"，就显示"欢迎你，管理员！"，否则显示"注册成功！"，并显示注册信息。

课 后 练 习

一、填空题

1. response 对象中用来动态改变 contentType 属性的方法是_____。
2. _____封装了属于客户会话的所有信息，该对象可以使用_____方法来设置指定名称的属性，使用_____方法可获得指定名称的属性。
3. request 对象可以使用_____方法获取表单提交的信息。
4. 客户端向服务器提交数据的方式通常有两种：一种是_____提交方式；另一种是_____提交方式。
5. out 内置对象中用来输出各种类型的数据并换行的方法是_____方法。
6. out 内置对象中用来输出缓冲区里的数据的方法是_____方法。

7. response 对象中用来把响应发送到另一个指定的位置进行处理的方法是_____方法。

8. _____对象可用于多个程序或者多个用户之间共享数据。

9. _____对象是从 JSP 脚本程序和表达式中获得的一个内置对象，它是 Object 类的一个实例。

二、选择题

1. 下列选项中，（　　）可以准确地获取请求页面的一个文本框的输入内容，文本框的名称为 name。

A. request.getParameter(name) 　　B. request.getParameter("name")
C. request.getParameterValues(name) 　　D. request.getParameterValues("name")

2. 在 response 对象进行重定向时，使用的是（　　）方法。

A. getAttribute 　B. setContentType 　C. sendRedirect 　D. setAttribute

3. 将 JSP 的内置对象按作用域由小到大排列的是（　　）。

A. request、application、session 　　B. session、request、application
C. request、session、application 　　D. application、request、session

4. （　　）内置对象可以处理 JSP 页面中的错误或异常。

A. pageContent 　B. page 　C. session 　D. exception

5. HttpServletRequest 接口的（　　）方法可用于创建会话。

A. setSession() 　B. getContext() 　C. getSession() 　D. putSession()

6. （　　）动作可用于将请求发送给其他页面。

A. next 　B. forward 　C. include 　D. param

7. page 指令的（　　）属性可用于引用需要的包或类。

A. extends 　B. import 　C. isErrorPage 　D. language

8. 假设 JSP 使用的表单中有如下的 GUI（复选框）：

```
<input type="checkbox"name="item" value="bird">鸟
<input type="checkbox"name="item" value="apple">苹果
<input type="checkbox"name="item" value="cat">猫
<input type="checkbox"name="item" value="moon">月亮
```

该表单请求的 JSP 可以使用 request 内置对象获取该表单提交的数据，那么下列哪个选项是 request 内置对象获取该表单提交的数据的正确语句？（　　）

A. String a=request.getParameter（"item"）；
B. String b=request.getParameter（"checkbox"）
C. String c[]=request.getParamterValues（"item"）
D. String d[]=request.getParameterValues（"checkbox"）

三、编程题

1. 用 out 内置对象在浏览器中输出服务器的系统时间，效果如图 3-30 所示。

图 3-30　输出服务器的系统时间的效果

2. 用 getServletInfo() 方法获取当前页面的 info 属性，效果如图 3-31 所示。

图 3-31　获取当前页面的 info 属性的效果

3. 用 response 对象编写一个自动刷新页面，实现每隔两秒钟自动刷新一次，刷新后的界面如图 3-32 所示。

图 3-32　刷新后的界面

4. 使用 session 对象制作站点计数器，效果如图 3-33 所示。

图 3-33　使用 session 对象制作站点计数器的效果

5. 使用 application 对象制作站点计数器，效果如图 3-34 所示。

图 3-34　使用 application 对象制作站点计数器的效果

项目四　JDBC 与数据库访问

项目要求

本项目是 JDBC 与数据库访问的应用，主要实现数据库的增、删、改、查功能，并应用数据库的相关知识完成真实项目中的数据库操作。

项目任务

要完成项目任务，至少需要具备三个基本条件：一是正确配置 MySQL 数据库，二是正确配置 JDBC 对应的 jar 包，三是掌握基本的驱动方法。该项目分为 4 个任务：Java 程序连接数据库、数据库查询和模糊查询、数据库更新、应用数据库连接池驱动 MySQL 数据库。

项目目标

【知识目标】掌握 JDBC 的概念；掌握 JDBC 常用的 API 的主要内容；掌握 Statement 接口和 Result 接口；熟悉 PreparedStatement 对象连接池。

【能力目标】能正确配置 jar 包；能应用 JDBC 实现数据库的增、删、改、查功能；能应用数据库连接池 Druid。

【素质目标】培养学生的网络安全意识，提升学生的民族自豪感和爱国热情。

知识导图

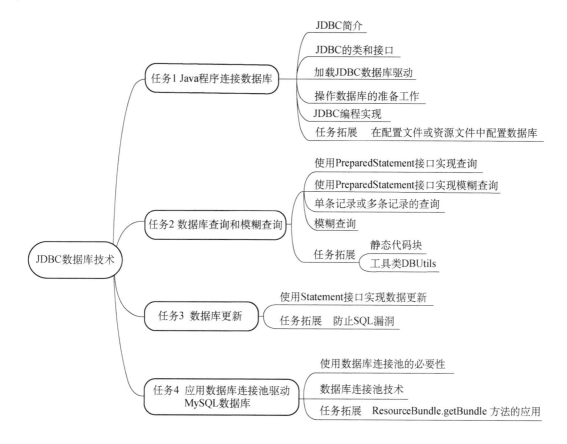

任务 1　Java 程序连接数据库

任务演示

本任务使用 Java 程序驱动 MySQL 数据库，将名为"jiaqicms"的数据库中的 user 表的记录数（即行数）打印到控制台上，效果如图 4-1 所示。

图 4-1　打印记录数的效果

注：此处输出的表名为"jq_user"，与操作的表名不一致，具体原因可参见后文的"教你一招"。

知识准备

1. JDBC 简介

JDBC（Java DataBase Connectivity，Java 数据库连接）提供了一种与平台无关的、用于

执行 SQL 语句的标准：Java API，可以方便用户实现多种关系型数据库的统一操作，它是一组用 Java 语言编写的类和接口。

在实际开发中不仅可以直接使用 JDBC 进行数据库的连接与操作，而且可以向数据库发送 SQL 命令。JDBC 提供了一套标准接口，只要根据提供的接口即可使用 JDBC 进行操作，极大地体现了 Java 语言"一次开发，多平台运行"的特点。

2. JDBC 的类和接口

JDBC 通过对特定厂商的数据库操作细节进行抽象，得到了一些类和接口，这些类和接口包含在 java.sql 包中，这样就可以被任何具有 JDBC 驱动的数据库使用，从而实现数据库访问的通用化。要使用 JDBC 连接不同厂商的数据库，首先应该安装对应厂商的数据库驱动。如果是 MySQL 数据库，只需安装 MySQL 驱动；如果是 Oracle 数据库，只需安装 Oracle 驱动；如果是 Access 数据库，只需安装 Access 驱动。

（1）Driver 接口

每个数据库驱动程序都必须实现 Driver 接口，代码如下。

```
Class.forName("com.mysql.jdbc.Driver");
```

其中，Class.forName 可用来对类进行初始化，它会让 Java 虚拟机（Java Virtual Machine，JVM）对指定的类进行加载、连接、初始化等操作。JVM 是专门为执行单个计算机程序而设计的，主要用来执行 Java 字节码指令。JVM 会查找指定路径类的 class 文件并将其读入内存，生成一个 class 对象作为访问数据库的入口，通过 getClass()方法可以获取 class 对象。

该接口可为类的类变量分配空间并赋值，执行静态代码块中的数据库的配置信息，实现对类的初始化操作。

Class.forName()方法的具体用法如下所示。

```
Class.forName("com.mysql.jdbc.Driver");
```

（2）DriverManager 类

DriverManager 类主要用来管理和注册数据库驱动，以及得到数据库连接对象。DriverManager 类定义了三个重要的静态方法，如表 4-1 所示。

表 4-1 DriverManager 类的静态方法

方法名称	作用
static registerDriver (Driver driver)	该方法用于在 DriverManager 类中注册给定的数据库驱动程序
static connection getConnection (String url,String user,String password)	创建连接对象，参数 url 为数据库服务器的地址，参数 user 为数据库用户名，参数 password 为用户的密码
static Driver getDriver (String url)	用于返回能够打开参数 url 指定的数据库驱动程序

【例 4-1】实现使用 Java 程序连接数据库。

```
public class MysqlDemo {
    public static void main(String[] args) {
        try {
            Class.forName("com.mysql.jdbc.Driver"); //1.注册驱动
            //2.获取连接
```

教学视频

```
            Connection  conn=DriverManager.getConnection("jdbc:mysql://localhost:3306/jiaqicms",
"root","root");
            System.out.println(conn);
        } catch (Exception e) {
            e.printStackTrace();
        }
    }
}
```

程序的运行结果如图 4-2 所示。

```
com.mysql.jdbc.JDBC4Connection@64616ca2
```

图 4-2　实现使用 Java 程序连接数据库的运行结果

（3）Connection 接口

Connection 接口代表 Java 程序和数据库之间的连接，只有建立连接才能访问数据库，实现对数据库的增、删、改、查等操作。该接口提供了多个方法，其中的 4 个常用方法如表 4-2 所示。

表 4-2　Connection 接口提供的常用方法

方法名称	作用
getMetaData()	该方法用于返回数据库的元数据的 DatabaseMetaData 对象
createStatement()	该方法用于创建一个 Statement 对象，并将 SQL 语句发送到数据库
prepareStatement (String sql)	该方法用于创建一个 prepareStatement 对象，并将参数化的 SQL 语句发送到数据库
prepareCall (String sql)	该方法用于创建一个 CallableStatement 对象来调用数据库的存储过程

【例 4-2】实例化 Statement 对象，代码如下。

```
public static void main(String[] args) {
    try {
        Class.forName("com.mysql.jdbc.Driver");//1.注册驱动
        //2.获取连接
        Connection  conn=DriverManager.getConnection("jdbc:mysql://localhost:3306/jiaqicms","root","root");
        Statement stmt = conn.createStatement(); //3.实例化 Statement 对象
        System.out.println(stmt);
    } catch (Exception e) {
        e.printStackTrace(); }
}}
```

教学视频

运行结果如图 4-3 所示。

```
<terminated> MysqlDemo [Java Application] D:\JSP\eclipse-
com.mysql.jdbc.StatementImpl@13e39c73
```

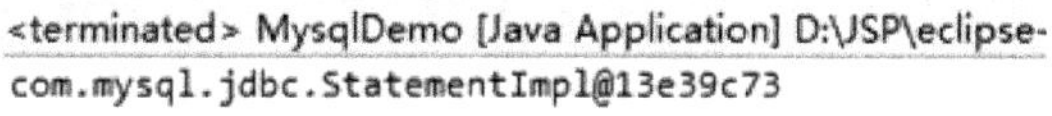

图 4-3　实例化 Statement 对象的运行结果

（4）Statement 接口

Statement 接口提供的方法如表 4-3 所示。

表 4-3　Statement 接口提供的方法

方法名称	作用
execute (String sql)	该方法可用于执行各种 SQL 语句，并返回一个 boolean 类型的值，如果返回值为 True，则表示执行的 SQL 语句有查询结果，且可通过 Statement 接口的 getResultSet()方法获得查询结果
executeUpdate (String sql)	该方法用于执行 insert、update 和 delete 语句。该方法返回一个 int 类型的值，表示数据库中受到该 SQL 语句影响的记录数
executeQuery (String sql)	该方法用于执行 select 语句，并返回一个表示查询结果的 ResultSet 对象

【例 4-3】Statement 接口和 executeQuery()方法的应用实例如下。

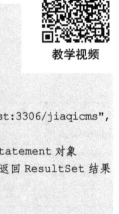

教学视频

```
public class MysqlDemo {
    public static void main(String[] args) {
        String sql="select count(*) from user";
        int count=0;
        try {
            Class.forName("com.mysql.jdbc.Driver");
Connection  conn=DriverManager.getConnection("jdbc:mysql://localhost:3306/jiaqicms",
"root","root");
            Statement stmt = conn.createStatement(); //实例化 Statement 对象
            ResultSet  rs=stmt.executeQuery(sql);//执行 SQL 语句,返回 ResultSet 结果
集对象

            System.out.println(rs);
        } catch (Exception e) {
        e.printStackTrace();
        }
    }
}
```

程序运行结果如图 4-4 所示。

```
<terminated> MysqlDemo [Java Application] D:\JSP\eclipse
com.mysql.jdbc.JDBC42ResultSet@7e0b85f9
```

图 4-4　Statement 接口和 executeQuery()方法的运行结果

（5）PreparedStatement 接口

Statement 接口封装了 JDBC 执行 SQL 语句的方法，它可以完成使用 Java 程序执行 SQL 语句的操作。在实际开发中往往需要将程序中的变量作为 SQL 语句的查询条件，而使用 Statement 接口操作带条件的 SQL 语句比较烦琐，因此出现了 PreparedStatement 接口。

PreparedStatement 接口是 Statement 接口的子接口，用于执行预编译的 SQL 语句。该接口扩展了带有参数的 SQL 语句的执行操作，先通过使用该接口中"？"来代替某个参数，

然后通过 setXxx()方法为 SQL 语句的参数指定值。PreparedStatement 接口提供了一些常用的方法，具体如表 4-4 所示。

表 4-4 PreparedStatement 接口提供的常用方法

方法名称	作用
executeUpdate(String sql)	该方法用于执行 update、insert、delete 语句，返回值是上述语句被执行后受到影响的记录数
executeQuery(String sql)	该方法用于执行 select 语句，返回值是查询得到的 ResultSet 对象结果集
setInt(int parameterIndex,int x)	该方法用于将给定的 int 类型的值赋给指定参数，parameterIndex 表示 SQL 语句的参数的位置
setFloat(int parameterIndex,float x)	该方法用于将给定的 float 类型的值赋给指定参数
setString(int parameterIndex,String x)	该方法用于将给定的 String 类型的值赋给指定参数
setDate(int parameterIndex,Date x)	该方法用于将给定的 Date 类型的值赋给指定参数

【例 4-4】PreparedStatement 接口提供的方法的应用实例如下。

```
public class MysqlDemo01 {
    public static void main(String[] args) {
        String sql="select * from jq_user where `username`=? and `password`=?";
        try {
            //1.注册驱动
            Class.forName("com.mysql.jdbc.Driver");
            //2.获取连接
            Connection conn=DriverManager.getConnection("jdbc:mysql://localhost:3306/jiaqicms","root","root");
            //3.实例化 PreparedStatement 对象,并进行预编译
            PreparedStatement pre = conn.prepareStatement(sql);
            String username="zhangsan";
            String password="123";
            pre.setString(1,username);
            pre.setString(2,password);
            //4. 执行 SQL 语句,返回 ResultSet 结果集对象
            ResultSet rs=pre.executeQuery();
            System.out.println(sql);
            System.out.println(rs);
        } catch (Exception e) {
            e.printStackTrace();
        }
    }
}
```

运行结果如图 4-5 所示。

```
select * from jq_user where `username`=? and `password`=?
com.mysql.jdbc.JDBC42ResultSet@3567135c
```

图 4-5 PreparedStatement 接口提供的方法的运行结果

（6）ResultSet 接口

在 ResultSet 接口内部有一个指向表格数据行的游标，当初始化 ResultSet 对象时，游标在表格的第 1 行之前，可调用 next()方法将游标移动到下一行。如果下一行没有数据，则返回的 boolean 类型的值为 False。ResultSet 接口中的常用方法如表 4-5 所示。

表 4-5　ResultSet 接口中的常用方法

方法名称	作用
getString(int columnIndex)	该方法用于获取指定字段的 String 类型的值，参数 columnIndex 代表字段的索引
getString(String columnName)	该方法用于获取指定字段的 String 类型的值，参数 columnName 代表字段的名称
getInt(int columnIndex)	该方法用于获取指定字段的 int 类型的值，参数 columnIndex 代表字段的索引
getInt(String columnName)	该方法用于获取指定字段的 int 类型的值，参数 columnName 代表字段的名称
setDate(int columnIndex)	该方法用于获取指定字段的日期类型的值，参数 columnIndex 代表字段的索引
setDate(String columnName)	该方法用于获取指定字段的日期类型的值，参数 columnName 代表字段的名称
next()	将游标从当前位置向下移一行
absolute(int row)	将游标移动到指定的行
afterLast()	将游标移动到指定的行末尾，即最后一行之后
beforeFirst()	将游标移动到指定的行开头，即第一行之前
Previous()	将游标移动到上一行
Last()	将游标移动到最后一行

从表 4-5 中可以看出，ResultSet 接口定义了大量的 getXxx()方法（Xxx 是方法名的一部分，具有见名知意的作用），而采用哪种 getXxx()方法取决于字段的数据类型。程序既可以通过字段的名称来获取指定的数据，也可以通过字段的索引来获取指定的数据，字段的索引是从 1 开始编号的。例如，数据表的第 1 列字段名为 id，字段的数据类型为 int 类型，那么既可以使用"getInt(1)"的字段索引方式获取该列的值，也可以使用"getInt("id")"的字段名称方式获取该列的值。

【脚下留心】
注意：ResultSet 接口中的值都可以通过 getString()方法获取，并且以字符串的形式按列编号或按列名称获取。

【例 4-5】next()方法和 getXxx()方法的应用实例如下。

```java
public class MysqlDemo04 {
    public static void main(String[] args) {
        String sql="select * from jq_user where `username`=? and `password`=?";
        try {
            //1.注册驱动
            Class.forName("com.mysql.jdbc.Driver");
            //2.获取连接
            Connection    conn=DriverManager.getConnection("jdbc:mysql://localhost:3306/jiaqicms","root","root");
            //3.实例化 PreparedStatement 对象,并进行预编译
            PreparedStatement pre = conn.prepareStatement(sql);
```

```
                String username="zhangsan";
                String password="123";
                pre.setString(1,username);
                pre.setString(2,password);
                //4. 执行 SQL 语句,返回 ResultSet 结果集对象
                ResultSet rs=pre.executeQuery();
                while(rs.next()) {
                System.out.println("获取到的姓名为:"+rs.getString("username"));
                }
        } catch (Exception e) {
            e.printStackTrace();
        }
    }}
```

运行程序，结果如图 4-6 所示。

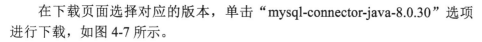

图 4-6　next()方法和 getXxx()方法的运行结果

从上面几个例子总结出使用 JDBC 编程主要有以下六个步骤。
第一步：注册数据库驱动，用于告诉 Java 程序即将连接的是哪个厂商的数据库。
第二步：获取数据库连接对象。
第三步：获取数据库操作对象，该对象主要用来执行 SQL 语句。
第四步：执行 SQL 语句。
第五步：处理查询到的结果集。
第六步：释放资源。

任务实施

这里使用的 MySQL 数据库驱动的 jar 包是 mysql-connector-java-8.0.30.jar，下面给出详细的任务实施步骤。

1. 加载 JDBC 数据库驱动

（1）下载 MySQL 数据库的 jar 包

MySQL 数据库所有版本的驱动都可以通过官方网站下载得到。

在下载页面选择对应的版本，单击"mysql-connector-java-8.0.30"选项进行下载，如图 4-7 所示。

教学视频

（2）解压下载文件

使用压缩软件解压"mysql-connector-java-8.0.30.zip"文件，得到解压文件 mysql-connector-java-8.0.30.jar，如图 4-8 所示。

（3）在 Eclipse 中使用 jar 包

在 Eclipse 中使用 jar 包，这里只介绍最常用的一种配置方法。

第一步：先新建 Java 项目，然后新建一个文件夹，文件夹名为"libs"，libs 文件夹主要用来存放外部包。右键依次选择工程名（这里单击的是工程"sqldemo1"）→"New"→"Folder"→"Folder name"→"libs"→"Finish"选项，如图 4-9 所示。

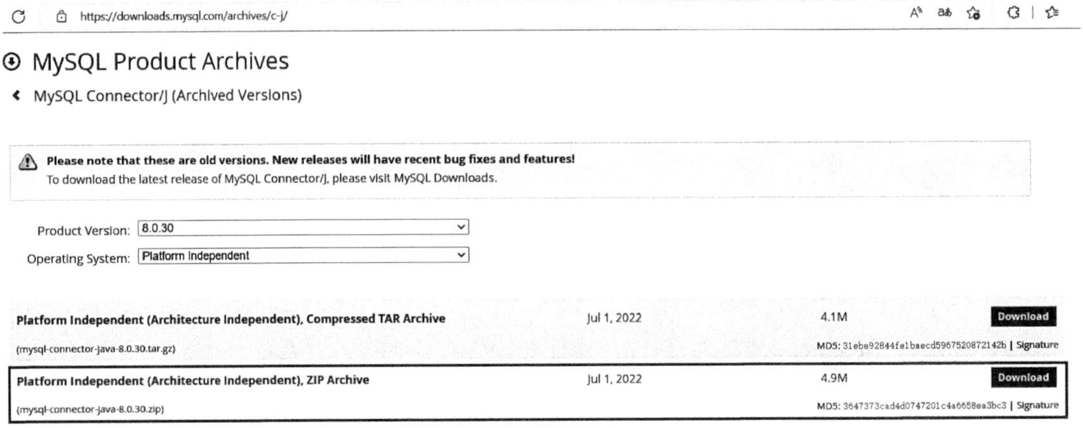

图 4-7　选择对应的版本进行下载

图 4-8　解压下载文件

图 4-9　新建 Java 项目和 libs 文件夹

第二步：在包里面加入连接 MySQL 数据库的包。

将刚才解压得到的 mysql-connector-java-8.0.30.jar 粘贴到 Java 项目的 libs 文件夹下（注意是物理地址）。此时，只需在 Eclipse 中右键单击 libs 文件夹进行刷新，就可将下载好的 JDBC 放到该文件夹下，如图 4-10 所示。

第三步：在 Eclipse 中依次单击"Project"→"Properties"按钮，如图 4-11 所示。

第四步：将 jar 包添加到项目中，操作流程如图 4-12 所示。

图 4-10　将 jar 包粘贴到文件夹下

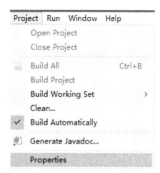
图 4-11　依次单击"Project"→"Porperties"按钮

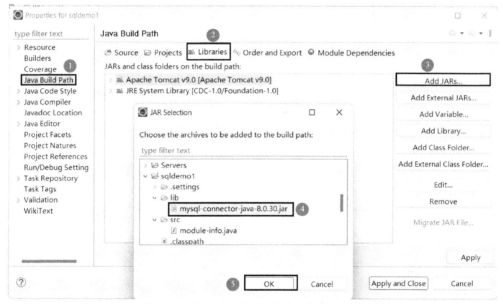
图 4-12　将 jar 包添加到项目中的操作流程

【教你一招】
注意：导入 jar 包有多种方法，最简单、直观的方法是先右键单击包名，然后依次选择"Build Path"→"Add to Build Path"选项即可。

2. 操作数据库的准备工作

要使用 JDBC 数据库驱动，并在 MySQL 服务器中创建数据库，需要使用 Navicat 创建一个名为"jiaqicms"的数据库，并在该数据库中创建一个名为"user"的表，具体步骤如下。

① 使用 Navicat 连接数据库，如图 4-13 所示。

② 新建数据库。先右键单击"本地数据库"文件夹，然后左键单击"新建数据库"选项，如图 4-14 所示。最后指定数据库名，设置编码，具体操作步骤如图 4-15 所示。

③ 新建表。先右键单击"表"选项，再单击"新建表"选项，如图 4-16 所示。

④ 创建名为"id"的字段，选择类型为"integer"，设置长度为 30，并且将其设置为主键，选中"自动递增"复选框，如图 4-17 所示。

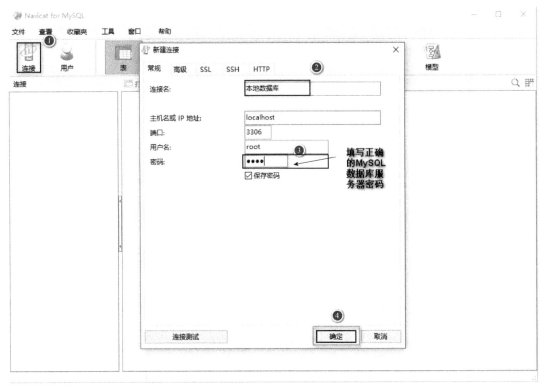

图 4-13 使用 Navicat 连接数据库

图 4-14 新建数据库

图 4-15　指定数据库名的具体操作步骤

图 4-16　新建表

⑤ 创建名为"username"的字段，按照如图 4-17 所示的步骤进行设置，得到新增的 username 字段如图 4-18 所示。

⑥ 通过单击"添加栏位"按钮增加"password""email""address"三个字段，如图 4-19 所示。

⑦ 单击"保存"按钮，在打开的"表名"对话框中输入表名"user"，单击"确定"按钮，如图 4-20 所示。

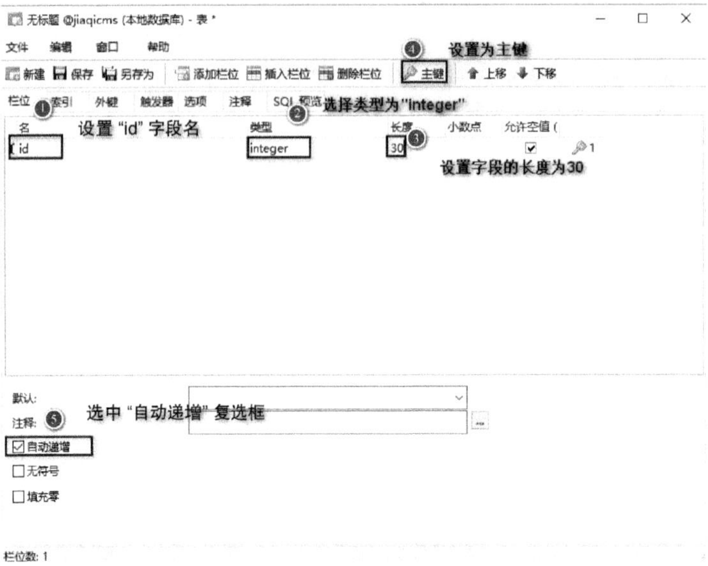

图 4-17 创建名为"id"的字段

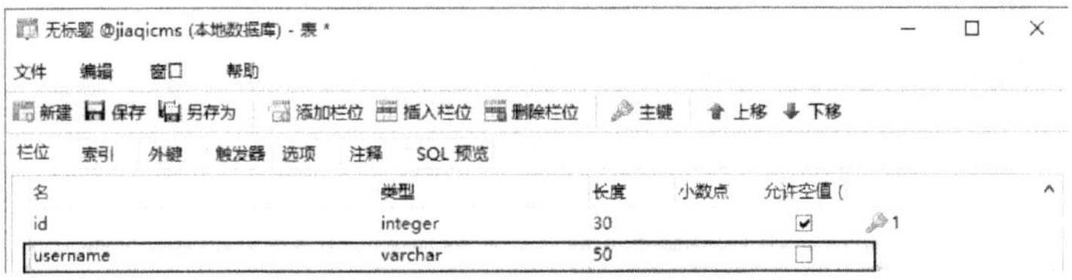

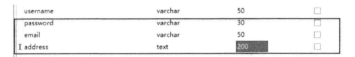

图 4-18 创建名为"username"的字段

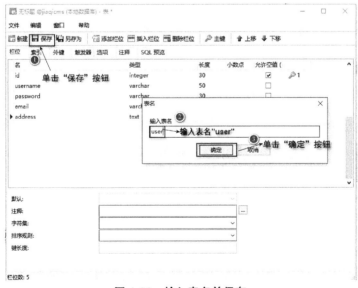

图 4-19 增加三个字段

图 4-20 输入表名并保存

⑧ 在表名保存成功后关闭"表名"对话框，双击 user 表图标，在打开的"user@jiaqicms（本地数据库）-表"对话框中输入相应的内容，如图 4-21 所示。

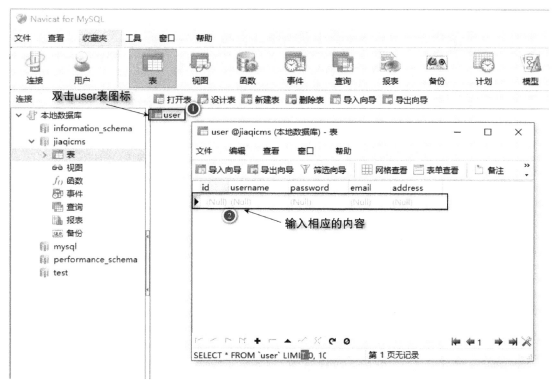

图 4-21　双击 user 表图标

⑨ 在 user 表中添加第一条记录，具体步骤如图 4-22 所示。

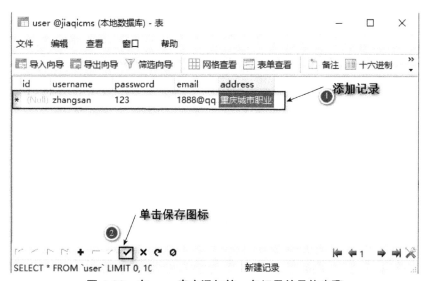

图 4-22　在 user 表中添加第一条记录的具体步骤

⑩ 继续新增记录，具体步骤如图 4-23 所示。
⑪ 保存新增记录的结果如图 4-24 所示。

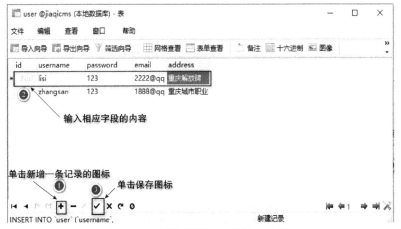

图 4-23　继续新增记录的具体步骤

图 4-24　保存新增记录的结果

【脚下留心】
　　注意：user 表中的 id 属于自动增长类型，id 在数据库中被设置为自动增长类型可以不用输入数据，且每条记录都会自动编号。

3. JDBC 编程实现

从前面所学的知识我们已经知道，要操作数据库，首先要加载数据库驱动，然后创建数据库连接，才能进行数据库的增、删、改、查操作。而且最后要关闭数据库连接，释放资源。下面将名为"jiaqicms"的数据库中的 jq_user 表（jq_user 表是 user 表的重命名表）的记录数查询出来，并打印到控制台上，代码如下，查询结果如图 4-25 所示。

```java
package cn.com.cqcvc;
import java.sql.Connection;
import java.sql.DriverManager;
import java.sql.ResultSet;
import java.sql.SQLException;
import java.sql.Statement;
public class MysqlDemo {
    public static void main(String[] args) {
        Connection conn=null;
        Statement stmt=null;
        String sql="select count(*) from jq_user";
        try {
            //1.注册驱动
```

```
            Class.forName("com.mysql.jdbc.Driver");
            //2.获取连接         conn=DriverManager.getConnection("jdbc:
mysql://localhost:3306/jiaqicms","root","root");
            //3.实例化对象
              stmt = conn.createStatement();
            // System.out.println(stmt);
            //4. 执行 SQL 语句,返回 ResultSet 结果集对象
            ResultSet rs=stmt.executeQuery(sql);
            while(rs.next()) {
             System.out.println("jq_user 表中一共有"+rs.getInt(1)+"行"记录);
            }
        } catch (Exception e) {
            e.printStackTrace();
        }
        try { conn.close();
            stmt.close();
        } catch (SQLException e) {
            e.printStackTrace();
        }
}}
```

```
<terminated> MysqlDemo [Java Application] D:\JSP\eclipse
jq_user表中一共有4行记录
```

图 4-25　查询结果

【教你一招】

注意：新建的表的表名为"user"，但"user"是 MySQL 数据库中的关键字，因此在对数据库进行增、删、改、查操作时会出错。这里有两种解决方法：第一种方法，使用命令"rename table user to jq_user;"修改表名，其中 user 是原先的表名，jq_user 是新的表名；第二种方法，不需要修改表名，而是给表名加上"''"，比如将"select * from user;"写成"select * from 'user';"。

任务拓展

在配置文件或资源文件中配置数据库

实际开发中，不建议在 Java 程序中把连接数据库的信息写为不可变的信息，而是建议将连接数据库的信息写入配置文件或资源文件中，这样方便修改相关文件，且不需要重新编译程序。

例如，在工程的 src 目录下创建一个名为"db.properties"的文件，并在该文件中输入如下内容：

教学视频

```
driver=com.mysql.jdbc.Driver
url=jdbc:mysql://localhost:3306/jiaqicms
username=root
password=root
```

【脚下留心】
注意：XX.properties 文件必须放在 src 目录下，不然会出错，而且编译器不会直接提示出错。

可以通过 ResourceBundle 类来读取后缀名为 ".properties" 的文件。ResourceBundle 类是 Java 自带的类，类路径为 java.util.ResourceBundle，主要用来读取项目中后缀名为 ".properties" 的文件。

可使用 getBundle() 方法获取该文件的名称，代码如下。

```
ResourceBundle resourceBundle = ResourceBundle.getBundle("db");
```

【教你一招】
注意：上面代码中的后缀名不需要在 getBundle() 方法中写出来。

可使用 getString() 方法获取资源文件中的信息，比如读取数据库的用户名的代码如下。

```
String username=bundle.getString("username");
```

在读取其他信息时，只需要改变 getString() 方法中的值就可以了。

将上述代码写到 db.properties 文件中，可得到 Java 部分的驱动代码如下。

```
package cn.com.cqcvc;
import java.sql.Connection;
import java.sql.DriverManager;
import java.sql.ResultSet;
import java.sql.SQLException;
import java.sql.Statement;
import java.util.ResourceBundle;
public class MysqlDemo {
    public static void main(String[] args) {
        Connection conn=null;
        Statement stmt=null;
        String sql="select count(*) from jq_user";
        String driver=null;
        String url=null;
        String username=null;
        String password=null;
        ResourceBundle bundle=ResourceBundle.getBundle("db");
        driver=bundle.getString("driver");
        url=bundle.getString("url");
        username=bundle.getString("username");
        password=bundle.getString("password");
```

```
            try {
                //1.注册驱动
                Class.forName(driver);
                //2.获取连接
                conn=DriverManager.getConnection(url,username,password);
                //3.实例化对象
                stmt = conn.createStatement();
                // System.out.println(stmt);
                //4. 执行 SQL 语句,返回 ResultSet 结果集对象
                ResultSet rs=stmt.executeQuery(sql);
                while(rs.next()) {
                    System.out.println("jq_user 表中一共有"+rs.getInt(1)+"行");
                }
            } catch (Exception e) {
                e.printStackTrace();
            }       try {
                conn.close();
                stmt.close();
            } catch (SQLException e) {
                
                e.printStackTrace();
            }
        }
    }
```

这里不再给出运行结果,请自行调试代码查看运行结果。

任务 2 数据库查询和模糊查询

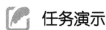任务演示

查询 jiaqicms 数据库中的 jq_user 表的第一条记录,并将该记录中的姓名和邮箱输出到控制台上,输出结果如图 4-26 所示。

教学视频

图 4-26 查询记录并输出到控制台上的输出结果

使用关键字 san 对 jiaqicms 数据库进行模糊查询,将查询到的记录中的"uername"字段的内容输出到控制台上,输出结果如图 4-27 所示,可以发现模糊查询的输出结果与上文的输出结果相同。

```
<terminated> MysqlDemo05 [Java Application] D:\JSP\eclipse
姓名：zhangsan----邮箱：1888@qq.com
```

图 4-27　模糊查询的输出结果

知识准备

前面已经实现了数据库的访问，访问到了数据库中的信息，并可以将信息输出到控制台上。下面我们使用 PreparedStatement 接口来实现数据库的查询和模糊查询。

1. 使用 PreparedStatement 接口实现查询

PreparedStatement 接口是 Statement 接口的子接口，属于预处理操作。与直接使用 Statement 接口不同的是，PreparedStatement 接口在操作时预先在数据表中准备了一条 SQL 语句，但是此 SQL 语句的具体内容暂时不设置，而是之后再进行设置。

使用 PreparedStatement 接口执行 SQL 语句与之前并没有什么不同，但对于具体的内容是使用"?"的占位符形式出现的，设置时要按照"?"的顺序设置具体的内容。定义 SQL 语句的示例代码如下。

```
String sql="select * from jq_user where username=?";
```

那么如何给"?"处的内容赋值呢？可以通过如下代码实现。

```
PreparedStatement stmt = conn.prepareStatement(sql);
stmt.setString(1,"zhangsan");
```

其中，"1"表示第一个"?"出现的位置，赋值内容为"zhangsan"。

2. 使用 PreparedStatement 接口实现模糊查询

完成查询操作以后，下面介绍模糊查询。模糊查询是利用数据的部分信息进行信息查找的一种查询方式。如果用户在进行数据查询时，不知道查询实体的具体信息，仅知道少量信息，此时就要用到模糊查询。下面我们首先介绍 MySQL 中的通配符的概念。

通配符是一种特殊语句，用来实现模糊查询。通配符主要有星号（*）和问号（?），可以用来代替一个或多个真正的字符。MySQL 中规定的通配符主要有以下三种。

① %：表示任何字符可以出现任意次（可以是 0 次）。

② _：只能匹配单个字符，不能多也不能少。

③ like：指定搜索模式是利用通配符进行比较的搜索模式，而不是直接相等匹配。如果 like 后面没有出现通配符，则在 SQL 执行优化时，会将 like 当作"="执行。

【例 4-6】对名为"jq_user"的表进行模糊查询。

```
String sql = "select username,password,email,address from jq_user" + " where username like ? ";
pstmt = conn.prepareStatement(sql);
pstmt.setString(1,"%" + username + "%"); // 设定参数
rs = pstmt.executeQuery();// 获取查询的结果集
```

任务实现

1. 单条记录或多条记录的查询

下面将 jq_user 表中姓名为"zhangsan"的记录打印到控制台上,如果 jq_user 表中有多条记录,则都打印到控制台上。

```java
package cn.com.cqcvc;
import java.sql.Connection;
import java.sql.DriverManager;
import java.sql.PreparedStatement;
import java.sql.ResultSet;
import java.sql.SQLException;
import java.sql.Statement;
import java.util.ResourceBundle;
public class MysqlDemo05 {
    public static void main(String[] args) {
        Connection conn=null;
        PreparedStatement stmt=null;
        String sql="select * from jq_user where username=?";
        String driver=null;
        String url=null;
        String username=null;
        String password=null;
        ResourceBundle bundle=ResourceBundle.getBundle("db");
        driver=bundle.getString("driver");
        url=bundle.getString("url");
        username=bundle.getString("username");
        password=bundle.getString("password");
        try {
            //1.注册驱动
            Class.forName(driver);
            //2.获取连接
            conn=DriverManager.getConnection(url,username,password);
            //3.实例化对象
            stmt = conn.prepareStatement(sql);
            stmt.setString(1,"zhangsan");

            //4. 执行 SQL 语句,返回 ResultSet 结果集对象
            ResultSet rs=stmt.executeQuery();
            while(rs.next()) {
                System.out.println("姓名:"+rs.getString("username")+"----"+"邮箱:"+rs.getString("email"));
            }
        } catch (Exception e) {
            e.printStackTrace();
        }
        try {
```

```
            conn.close();
            stmt.close();
        } catch (SQLException e) {

            e.printStackTrace();
        }
    }
}
```

查询结果如图 4-28 所示,因 user 表中只有一条记录,所以查询结果只显示了一条记录。

图 4-28 单条记录或多条记录的查询结果

2. 模糊查询

模糊查询主要用于网站内容的搜索,使用下列代码实现对 jq_user 表的模糊查询。

```java
package cn.com.cqcvc;
import java.sql.Connection;
import java.sql.DriverManager;
import java.sql.PreparedStatement;
import java.sql.ResultSet;
public class MysqlDemo06 {
    public static void main(String[] args) {
        String sql="select * from jq_user where username like ?";
        try {
            Class.forName("com.mysql.jdbc.Driver");  //1.注册驱动
            //2.获取连接
Connection  conn=DriverManager.getConnection("jdbc:mysql://localhost:3306/jiaqicms","root","root")
   PreparedStatement pre = conn.prepareStatement(sql);//3.实例化对象,进行预编译
    String username="%san%";
    pre.setString(1,username);                  //4. 执行SQL语句,返回ResultSet结果集对象
    ResultSet rs=pre.executeQuery();
            while(rs.next()) {
            System.out.println("获取到的姓名为:"+rs.getString("username"));
            }
        } catch (Exception e) {
            e.printStackTrace();
        }
    }
}
```

程序运行结果如图 4-29 所示。

```
<terminated> MysqlDemo06 [Java Application] D:\JSP\eclipse
获取到的姓名为：zhangsan
```

图 4-29　模糊查询的程序运行结果

任务拓展

1. 静态代码块

在使用 JDBC 操作数据库时，为了增加代码的通用性，可以将数据库连接等操作抽象为一个工具类 DBUtils，这需要用到静态代码块的知识。那么什么是静态代码块呢？

静态代码块是定义在类中，使用关键字 static 修饰的代码块。无论类在哪个位置，不论类产生多少个对象，静态代码块只在加载类时被执行一次，如下是静态代码块的应用。

```java
package cn.com.cqcvc;
public class testDemo02 {
    public static void main(String[] args) {
        System.out.println("进入了main方法！");
        Person p=new Person("张三","123");
    }
}
class Person{
    String username;
    String password;
    public Person() {
        System.out.println("这是无参构造函数！");
    }
    public Person(String username,String password) {
    super();
    this.username = username;
    this.password = password;
        System.out.println("这是有参构造函数！");
    }
    static {
        System.out.println("这是静态代码块！");
    }
}
```

运行程序，结果如图 4-30 所示。

```
<terminated> testDemo02 [Java Application] D:\JSP\eclipse
进入了main方法！
这是静态代码块！
这是有参构造函数！
```

图 4-30　静态代码块只在加载类时被执行一次的结果

2. 工具类 DBUtils

在掌握了静态代码块的知识后，尝试把数据库的操作做成一个工具类 DBUtils，代码如下。

教学视频

```java
package cn.com.cqcvc;
import java.sql.Connection;
import java.sql.DriverManager;
import java.sql.PreparedStatement;
import java.sql.ResultSet;
import java.util.ResourceBundle;
public class DBUtils {
    private static Connection conn=null;
    private static String driver=null;
    private static String url=null;
    private static String username=null;
    private static String password=null;
public static void readFile() throws Exception {
    ResourceBundle bundle=ResourceBundle.getBundle("db");
     driver=bundle.getString("driver");
    url=bundle.getString("url");
    username=bundle.getString("username");
    password=bundle.getString("password");
}
static {
    //只加载一次驱动
    try {
        readFile();//读取本地配置文件
        Class.forName(driver);
    } catch (Exception e) {
        e.printStackTrace();
    }
}
    //获取连接
    public static Connection getConnection() throws Exception {
        conn = DriverManager.getConnection(url,username,password);
        return conn;
    }
    public static void myclose(Connection conn,ResultSet rs,PreparedStatement pmst) throws Exception{
            if(conn!=null) {
                conn.close();
            }
            if(rs!=null) {
                rs.close();
            }
            if(pmst!=null) {
                pmst.close();
```

```
        }
    }
}
```

在有了工具类 DBUtils 以后，还需要编写一个测试方法，代码如下。

```
package cn.com.cqcvc;
import java.sql.Connection;
import java.sql.PreparedStatement;
import java.sql.ResultSet;
public class MysqlDemo10 {
    public static void main(String[] args) throws Exception{
        DBUtils db=new DBUtils();
        Connection conn=db.getConnection();
        String sql="select * from jq_user";
        PreparedStatement pmst =   conn.prepareStatement(sql);
        ResultSet rs=pmst.executeQuery();
        while(rs.next()) {
            System.out.println("用户名:"+rs.getString("username")+",邮箱:"+rs.getString("email"));
        }
        db.myclose(conn,rs,pmst);
    }
}
```

运行程序，结果如图 4-31 所示。

```
<terminated> MysqlDemo10 [Java Application] D:\JSP\eclipse
用户名: zhangsan,邮箱: 1888@qq.com
用户名: lisi,邮箱: 2222@qq.com
用户名: klkl,邮箱: kkk@jjj
用户名: lier,邮箱: 233@qq.com
```

图 4-31　编写一个测试方法的结果

任务 3　数据库更新

在连接数据库以后，就可以对数据库进行增、删、改、查等操作了。可以使用 Statement 接口完成数据更新的操作，该接口需要 Connection 接口提供的 creatStatement()方法进行实例化，进而实现相应的操作。

 任务演示

本任务使用 Java 程序驱动 MySQL 数据库，在 jiaqicms 数据库中的 jq_user 表中增加数据，并实现修改、删除的操作，操作界面如图 4-32 所示。

图 4-32 实现修改、删除的操作界面

知识准备

使用 Statement 接口实现数据更新

可以使用 Statement 接口实现数据更新，JDBC 中的 Statement 接口可用于向数据库发送 SQL 语句。要完成数据库的增、删、改、查操作，只需要通过这个接口向数据库发送增、删、改、查的 SQL 语句即可。

Statement 接口的 executeUpdate()方法可用于向数据库发送增、删、改、查的 SQL 语句，在 executeUpdate()方法执行完后，将会返回一个整数。

（1）插入数据

【例 4-7】使用 executeUpdate(String sql)方法添加数据，代码如下。

```
Statement stmt = conn.createStatement();
    String sql="insert into user(...) values(...)";
    int num=stmt.executeUpdate(sql);
    if(num>0) {
        System.out.println("数据插入成功");}
```

（2）删除数据

【例 4-8】使用 executeUpdate(String sql)方法将表中的指定数据删除，代码如下。

教学视频

```
Statement stmt = conn.createStatement();
       String sql="delete from jq_user where id=1";
       int num=stmt.executeUpdate(sql);
       if(num>0) {
           System.out.println("数据删除成功");
       }
```

（3）修改数据

【例 4-9】使用 executeUpdate(String sql)方法修改表中的指定数据。

```
Statement stmt = conn.createStatement();
      String sql="update jq_user set username='' where id=1";
      int num=stmt.executeUpdate(sql);
      if(num>0) {
          System.out.println("数据修改成功");
      }
```

教学视频

任务实施

① 在 jq_user 表中增加一条新的记录，编写并执行一条完整的 SQL 语句。

```java
package cn.com.cqcvc;
import java.sql.Connection;
import java.sql.DriverManager;
import java.sql.SQLException;
import java.sql.Statement;
public class MysqlDemo07 {
public static void main(String[] args) throws Exception{
    Connection conn=null;
    Statement stmt=null;
        Class.forName("com.mysql.jdbc.Driver");
        conn = DriverManager.getConnection("jdbc:mysql://localhost:3306/jiaqicms","root","root");
        stmt =  conn.createStatement();
    String sql="insert into jq_user(username,password,email,address) values('lier','123','233@qq.com','南京路')";
        int num=stmt.executeUpdate(sql);
        if(num>0) {
        System.out.println("数据增加成功");
        conn.close();
        stmt.close();
}
}
}
```

运行结果如图 4-33 所示。

id	username	password	email	address
1	zhangsan	123	1888@qq.com	重庆城市职业学院
2	lisi	123	2222@qq.com	重庆解放碑
3	klkl	123	kkk@jjj	南滨路
4	wangyong	123	233@qq.com	南京路
5	lier	123	233@qq.com	南京路

图 4-33　在 jq_user 表中添加一条新的记录的运行结果

② 将 jq_user 表中 id 字段的值为 "5" 的记录中 password 字段的值修改为 "123456"，同时将 address 字段的值修改为 "上海东路 100 号"，对应的程序代码如下。

```java
package cn.com.cqcvc;
import java.sql.Connection;
import java.sql.DriverManager;
import java.sql.Statement;
public class MysqlDemo08 {
    public static void main(String[] args) throws Exception {
        Connection conn=null;
        Statement stmt=null;
        Class.forName("com.mysql.jdbc.Driver");
        conn = DriverManager.getConnection("jdbc:mysql://localhost:3306/
```

```
jiaqicms","root","root");
                stmt  =  conn.createStatement();
                String sql="update  jq_user set `password`='123456',`address`='上海东路100号' where id=5";
                int num=stmt.executeUpdate(sql);
                if(num>0) {
                System.out.println("数据修改成功");
                conn.close();
                stmt.close();
        }
    }}
```

运行程序，结果如图 4-34 所示。

←T→				id	username	password	email	address
☐	编辑	复制	删除	1	zhangsan	123	1888@qq.com	重庆城市职业学院
☐	编辑	复制	删除	2	lisi	123	2222@qq.com	重庆解放碑
☐	编辑	复制	删除	3	klkl	123	kkk@jjj	南滨路
☐	编辑	复制	删除	4	wangyong	123	233@qq.com	南京路
☐	编辑	复制	删除	5	lier	123456	233@qq.com	上海东路100号

图 4-34　修改 jq_user 表的记录的结果

③ 将 jq_user 表中 id 字段的值为“4”的记录删除，对应的代码如下。

```
    package cn.com.cqcvc;
    import java.sql.Connection;
    import java.sql.DriverManager;
    import java.sql.Statement;
    public class MysqlDemo09 {
    public static void main(String[] args) throws Exception{
            Connection conn=null;
            Statement stmt=null;
                Class.forName("com.mysql.jdbc.Driver");
            conn = DriverManager.getConnection("jdbc:mysql://localhost:3306/jiaqicms",
"root","root");
                stmt  =  conn.createStatement();
                String sql="delete from jq_user where id=4";
                int num=stmt.executeUpdate(sql);
                if(num>0) {
                System.out.println("数据删除成功");
                conn.close();
                stmt.close();
        }
        }
    }
```

运行结果如图 4-35 所示。

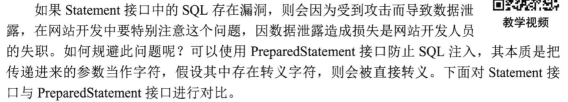

图 4-35　将 jq_user 表中的记录删除的运行结果

任务拓展

防止 SQL 漏洞

如果 Statement 接口中的 SQL 存在漏洞，则会因为受到攻击而导致数据泄露，在网站开发中要特别注意这个问题，因数据泄露造成损失是网站开发人员的失职。如何规避此问题呢？可以使用 PreparedStatement 接口防止 SQL 注入，其本质是把传递进来的参数当作字符，假设其中存在转义字符，则会被直接转义。下面对 Statement 接口与 PreparedStatement 接口进行对比。

① 在使用 Statement 接口登录时被 SQL 注入。

```java
package cn.com.cqcvc;
import java.sql.Connection;
import java.sql.DriverManager;
import java.sql.ResultSet;
import java.sql.Statement;
import java.util.Scanner;
public class loginDemo01 {
    public static void main(String[] args) throws Exception {
        Connection conn=null;
        Statement stmt=null;
        String username="zhangsan";
        String password="123";
        // 输入账号和密码
        Scanner sc = new Scanner(System.in);
        System.out.println("请输入账号:");
        username = sc.nextLine();
        System.out.println("请输入密码:");
        password = sc.nextLine();
        Class.forName("com.mysql.jdbc.Driver");
        conn = DriverManager.getConnection("jdbc:mysql://localhost:3306/jiaqicms","root","root");
        stmt = conn.createStatement();
        String sql = "select * from jq_user where username='" + username+ "' and password='" + password + "'";
        System.out.println(sql);
        ResultSet rs=stmt.executeQuery(sql);

        if(rs.next()) {
```

```
            System.out.println("登录成功");
        }else {
            System.out.println("登录失败");
        }
        conn.close();
        stmt.close();
    }
}
```

运行结果如图 4-36 所示。

```
<terminated> loginDemo01 [Java Application] D:\JSP\eclipse-2022-6\plugins\org.eclipse.justj.openjdk.hotspot.jre.full
请输入账号：
zhangsan
请输入密码：
1' or ' 1=1
select * from jq_user where username='zhangsan' and password='1' or ' 1=1
登录成功
```

图 4-36　在使用 Statement 接口登录时被 SQL 注入的运行结果

② 使用 PreparedStatement 接口防止 SQL 注入。

从 jq_user 表可知，password 字段的值没有 "1"，但是却登录成功了，这是 SQL 注入造成的。为了防止这种情况，可以使用 PreparedStatement 接口来防止 SQL 注入。

教学视频

```
package cn.com.cqcvc;
import java.sql.Connection;
import java.sql.DriverManager;
import java.sql.PreparedStatement;
import java.sql.ResultSet;
import java.sql.Statement;
import java.util.Scanner;
public class loginDemo02 {
public static void main(String[] args) throws Exception {
        Connection conn=null;
        PreparedStatement stmt=null;
        String username="zhangsan";
        String password="123";
        // 输入账号和密码
    Scanner sc = new Scanner(System.in);
    System.out.println("请输入账号:");
     username = sc.nextLine();
    System.out.println("请输入密码:");
    password = sc.nextLine();
            Class.forName("com.mysql.jdbc.Driver");
            String sql = "select * from jq_user where username=? and password=?";
            conn  =  DriverManager.getConnection("jdbc:mysql://localhost:3306/jiaqicms","root","root");
```

```
            stmt =   conn.prepareStatement(sql);
            stmt.setString(1,username);
            stmt.setString(2,password);
            ResultSet rs=stmt.executeQuery();
            System.out.println(sql);
            if(rs.next()) {
            System.out.println("登录成功");
            }else {
                System.out.println("登录失败");
            }
            conn.close();
            stmt.close();
    }   }
```

运行结果如图 4-37 所示。

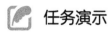

```
<terminated> loginDemo02 [Java Application] D:\JSP\eclipse-2022-6\plugins\org.eclipse.
请输入账号:
zhangsan
请输入密码:
1 ' or '1=1
select * from jq_user where username=? and password=?
登录失败
```

图 4-37　使用 PreparedStatement 接口防止 SQL 注入的运行结果

任务 4　应用数据库连接池驱动 MySQL 数据库

任务演示

本任务要使用 JDBC 的数据库连接池访问 MySQL 数据库，如图 4-38 所示。

```
12
13  public class DruidUtil  {
14      private static Connection conn=null;
15      private static DataSource dataSource;
16      static {
17
18  try {
19
20      InputStream is =
21      DruidUtil.class.getClassLoader().getResourceAsStream("druid.properties");
22      Properties prop = new Properties();
23      prop.load(is);
24      dataSource = DruidDataSourceFactory.createDataSource(prop);
25  } catch (Exception e) {
26      e.printStackTrace();
27  }
28      }
29
30  //获取连接
31  public static Connection getConnection() throws Exception {
32
33      Connection conn = dataSource.getConnection();
34
35      return conn;
36
37  }
```

图 4-38　使用 JDBC 的数据库连接池访问 MySQL 数据库

知识准备

1. 使用数据库连接池的必要性

在应用通用方法访问数据库时，主要包括以下三个步骤。

第一步：在主程序中建立数据库连接。

第二步：进行 SQL 操作。

第三步：断开数据库连接。

应用这种通用方法存在以下几个问题。

① 普通的 JDBC 数据库连接可使用 DriverManager 来获取，每次向数据库建立连接都要先将 Connection 加载到内存中，再验证用户名和密码（需花费 0.05～1s 的时间）。在需要连接数据库的时候，就向数据库发送请求，直到执行完成再断开连接。这样的方式会消耗大量的资源和时间，数据库的连接资源并没有得到很好的利用。若同时有几百人甚至几千人频繁进行数据库连接操作，将会占用很多的系统资源，甚至会造成服务器崩溃。

② 每一次数据库连接在使用完后都得断开，如果程序出现异常而未能断开数据库连接，将会导致数据库系统的内存泄漏，最终重启数据库。

③ 这种通用方法不能控制被创建的连接对象数，系统资源会被毫无顾忌地分配出去，如连接对象过多可能导致内存泄漏、服务器崩溃。

采用数据库连接池技术可以解决传统开发中存在的数据库连接问题。

2. 数据库连接池技术

数据库连接池技术的基本思想是为数据库连接建立一个"缓冲池"。预先在"缓冲池"中放入一定数量的连接，当需要建立数据库连接时，只需从"缓冲池"中取出，使用完毕再放回去。

数据库连接池负责建立、管理和释放数据库连接，它允许应用程序重复使用一个现有的数据库连接，而不是重新建立一个数据库连接。

在数据库连接池初始化时会在连接池中创建一定数量的数据库连接，这些数据库连接的数量是由最小数据库连接的数量来设定的。无论这些数据库连接是否被使用，连接池都将一直保证一定的连接数量。连接池的最大数据库连接数量限定了这个连接池能占有的最大连接数量，当应用程序向连接池请求的连接数量超过最大数据库连接数量时，这些请求将被加到等待队列中。数据库连接池示意图如图 4-39 所示。

Druid 连接池是阿里巴巴开源的数据库连接池。Druid 连接池为监控而生，因此内置了强大的监控功能，且不影响性能，能防止 SQL 注入，其内置 Loging 能诊断 Hack 应用行为。Druid 连接池号称业界最优秀的连接池，在性能、监控、诊断、安全、扩展性等方面的表现都非常不错。

在使用 Druid 连接池时，先使用 DruidDataSourceFactory 根据配置文件的内容创建出 DataSource 数据源对象，然后调用 getConnection()方法获取数据库连接对象，在得到数据库连接对象之后，其他的操作跟 JDBC 访问数据库的操作大致一样，唯一的区别就是当调用数据库连接对象的 close()方法时，底层不再关闭或销毁数据库连接对象，而是将数据库连接对象放入连接池中，以便后续有新的请求到来时，可以直接拿去使用。

在项目的 resources 文件夹下创建 Druid 连接池的配置文件，配置文件可以随意命名，但

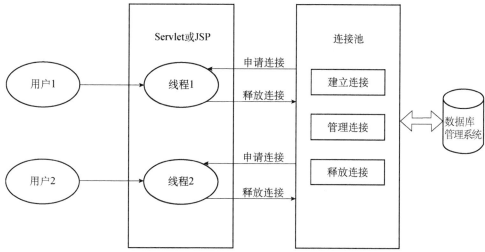

图 4-39 数据库连接池示意图

是文件内容中每项配置的 key 必须是指定的名称，这里使用"druid.properties"作为配置文件的名称，其内容如下所示。

```
# 数据库连接参数
driverClassName=com.mysql.jdbc.Driver
url=jdbc:mysql://localhost:3306/testdb
username=root
password=123456
# 初始化数据库连接的数量 initialSize=3
# 最大数据库连接数量 maxActive=20
# 获取数据库连接的最长等待时间(毫秒)
maxWait=3000
```

有了配置文件以后，就可以使用如下代码读取配置文件并得到数据源对象，通过数据源对象可获取数据库连接对象。

教学视频

```
//加载配置文件
    InputStream is = druidtest.class.getClassLoader().
getResourceAsStream("db.properties");
    Properties prop = new Properties();
    prop.load(is);
//根据配置文件内容,创建出数据源对象
    DataSource dataSource = DruidDataSourceFactory.createDataSource(prop);
    //通过数据源对象获取数据库连接对象
//如果连接池中的数据库连接已经被用完,则会等待一定时间(所配置的时间)
//如果等待超时,就会抛出异常
    Connection conn = dataSource.getConnection();
System.out.println(conn);
```

可见，有了数据库连接对象以后，就可以很方便地操作数据库了。

项目实施

第一步:创建名为"DruidUtil"的工具类,代码如下。

```java
package cn.com.cqcvc;
import java.io.InputStream;
import java.sql.Connection;
import java.sql.PreparedStatement;
import java.sql.ResultSet;
import java.util.Properties;
import javax.sql.DataSource;
import com.alibaba.druid.pool.DruidDataSourceFactory;
public class DruidUtil {
    private static Connection conn=null;
    private static DataSource dataSource;
    static {
            try {
                InputStream is =
                        DruidUtil.class.getClassLoader().getResourceAsStream("druid.properties");
                    Properties prop = new Properties();
                    prop.load(is);
                    dataSource = DruidDataSourceFactory.createDataSource (prop);
                } catch (Exception e) {
                    // TODO Auto-generated catch block
                    e.printStackTrace();
                }
        }
    //获取连接
    public static Connection getConnection() throws Exception {
        Connection conn = dataSource.getConnection();
        return conn;
    }
    public static void myclose(Connection conn,ResultSet rs,PreparedStatement pmst) throws Exception{
        if(conn!=null) {
            conn.close();
        }
        if(rs!=null) {
            rs.close();
        }
        if(pmst!=null) {
            pmst.close();
        }
    }
}
```

第二步：编写测试代码。

```java
package cn.com.cqcvc;
import java.sql.Connection;
import java.sql.PreparedStatement;
import java.sql.ResultSet;
public class MysqlDemo11 {
public static void main(String[] args) throws Exception{
        DruidUtil db=new DruidUtil();
        Connection conn=db.getConnection();
        String sql="select * from jq_user";
        PreparedStatement pmst = conn.prepareStatement(sql);
        ResultSet rs=pmst.executeQuery();
        while(rs.next()) {
        System.out.println("用户名:"+rs.getString("username")+",邮箱:"+rs.getString("email"));
            }
        db.myclose(conn,rs,pmst);
    }
}
```

运行程序，结果如图 4-40 所示。

图 4-40　使用 JDBC 的数据库连接池访问数据库的结果

任务拓展

ResourceBundle.getBundle 方法的应用

也可以使用 ResourceBundle.getBundle 方法读取配置文件，读取方法已经介绍过了，具体操作步骤如下。

教学视频

第一步：创建名为"DruidUtils"的工具类，代码如下。

```java
package cn.com.cqcvc;
import java.sql.Connection;
import java.sql.DriverManager;
import java.sql.PreparedStatement;
import java.sql.ResultSet;
import java.util.ResourceBundle;
import com.alibaba.druid.pool.DruidDataSource;
    public class DruidUtils {
            private static Connection conn=null;
```

```java
        private static String driver=null;
        private static String url=null;
        private static String username=null;
        private static String password=null;
    public static void readFile() throws Exception {
        ResourceBundle bundle=ResourceBundle.getBundle("db");
        driver=bundle.getString("driver");
        url=bundle.getString("url");
        username=bundle.getString("username");
        password=bundle.getString("password");
    }
        //获取连接
    public static Connection getConnection() throws Exception {
        readFile();
        DruidDataSource ds=new DruidDataSource();
        ds.setDriverClassName(driver);
        ds.setUrl(url);
        ds.setUsername(username);
        ds.setPassword(password);
        ds.setMaxActive(10);
        conn = ds.getConnection();
        return conn;
    }
    public static void myclose(Connection conn,ResultSet rs,PreparedStatement pmst) throws Exception{
        if(conn!=null) {
            conn.close();
        }
        if(rs!=null) {
            rs.close();
        }
        if(pmst!=null) {
            pmst.close();
        }
    }
    public static void main(String[] args) throws Exception {
        DruidUtils db=new DruidUtils();
        System.out.println(db.getConnection());
    }
}
```

第二步：编写测试方法。

```java
package cn.com.cqcvc;
import java.sql.Connection;
import java.sql.PreparedStatement;
import java.sql.ResultSet;
public class MysqlDemo10 {
    public static void main(String[] args) throws Exception{
        DruidUtils db=new DruidUtils();
```

```java
            Connection conn=db.getConnection();
            String sql="select * from jq_user";
            PreparedStatement pmst =  conn.prepareStatement(sql);
            ResultSet rs=pmst.executeQuery();
            while(rs.next()) {
                System.out.println("用户名:"+rs.getString("username")+",邮箱:"+rs.getString("email")+",地址:"+rs.getString("address"));
            }
            db.myclose(conn,rs,pmst);
    }
```

运行程序，结果如图 4-41 所示。

```
<terminated> MysqlDemo10 [Java Application] D:\JSP\eclipse-2022-6\plugins\org.eclipse.justj.op
7月 16, 2022 4:26:33 下午 com.alibaba.druid.pool.DruidDataSource
信息: {dataSource-1} inited
用户名:zhangsan,邮箱:1888@qq.com,地址:重庆城市职业学院
用户名:lisi,邮箱:2222@qq.com,地址:重庆解放碑
用户名:klkl,邮箱:kkk@jjj,地址:南滨路
用户名:lier,邮箱:233@qq.com,地址:上海东路100号
```

图 4-41 应用 ResourceBundle.getBundle 方法的结果

项 目 实 训

实训 创建数据库并进行相关的数据库操作

MySQL 数据库中的员工表（employee）如图 4-42 所示。

	EMP_ID	EMP_NAME	JOB	SALARY	DEPT
1	1	王楠	clerk	4300.00	10
2	2	张敏	clerk	4300.00	10
3	3	李刚	manager	5000.00	20
▶4	4	马明	manager	5000.00	20

图 4-42 员工表

要求：使用 JDBC 的数据库连接池技术编程实现下列功能。
① 添加员工信息到员工表中；
② 修改员工的基本信息；
③ 根据编号删除员工信息；
④ 按照员工的工作种类进行员工信息查询。

课 后 练 习

一、填空题

1. _____是一种用于执行 SQL 语句的 Java API，可为多种关系数据库提供统一的访问，由一组用 Java 语言编写的类和接口组成。

2. JDBC API 供程序员调用的接口与类，集成在_____和_____包中。

3. JDBC 可做三件事：_____、_____、_____。

4. 加载数据库驱动是通过调用_____方法实现的。

5. JDBC 与数据库建立连接是通过调用 DriverManager 类的静态方法_____实现的。

6. 有三种 Statement 对象，分别是_____、_____、_____。

7. _____对象是 executeQuery()方法的返回值，也被称为结果集，它代表符合 SQL 语句的所有行数据，并且通过 getXXX()方法（这些方法可以访问当前行的不同列数据）提供了对这些行数据的访问。

8. ResultSet 对象自动维护指向当前行数据的游标。每调用一次_____方法，游标向下移动一行。

9. 在 JDBC 中，如果事务操作成功，系统将自动调用_____提交，否则调用_____回滚。

10. 在 JDBC 中，事务操作方法都位于 java.sql.Connection 接口中，可以通过调用_____来禁止自动提交。

11. 在 JDBC 中，事务开始的边界不明显，它开始于组成当前事务的第一个_____被执行的时候。

二、选择题

1. 以下选项中，有关 Connection 对象的描述错误的是（   ）。

A. Connection 对象是 Java 程序与数据库的连接对象，只能用来连接数据库，不能执行 SQL 语句

B. JDBC 的数据库事务控制要靠 Connection 对象完成

C. Connection 对象使用完毕后要及时关闭，否则会对数据库造成负担

D. 只有 MySQL 数据库和 Oracle 数据库的 JDBC 程序需要创建 Connection 对象才会执行 CRUD 操作，其他数据库的 JDBC 程序不用创建 Connection 对象就可以执行 CRUD 操作

2. 使用 Connection 对象的哪个方法可以建立一个 PreparedStatement 接口？（   ）

A. createPrepareStatement()　　　　　B. prepareStatement()

C. createPreparedStatement()　　　　 D. preparedStatement()

3. 在 JDBC 编程中，执行完 SQL 语句"SELECT name,rank,serialNo FROM employee"，能得到 rs 的第一列数据的代码是（   ）。

A. rs.getString(0);　　　　　　　　　B. rs.getString("name");

C. rs.getString(2);　　　　　　　　　D. rs.last();

项目五　JavaBean 开发模型

项目要求

本项目是 JavaBean 开发模型的应用，主要完成 JavaBean 模型的创建和应用，提高程序的可读性和重用性，节省开发时间。

项目分析

要完成项目任务，至少需要具备两个基本条件：一是掌握 JavaBean 的创建方法，二是掌握 JavaBean 的综合应用。本项目分为 3 个任务，分别是应用 JavaBean 计算梯形的面积、应用 JavaBean 实现化妆品网站注册功能和应用 JavaBean 实现水果购物车系统。

项目目标

【知识目标】掌握 JSP 中编写 JavaBean 的方法；掌握 JavaBean 数据库访问方法和编码转换方法。

【能力目标】能编写和应用 JavaBean；能应用 JavaBean 实现数据库的封装，并解决中文乱码问题。

【素质目标】培养学生编写代码的规范性和严谨性；提高学生分析问题、解决问题的能力。

知识导图

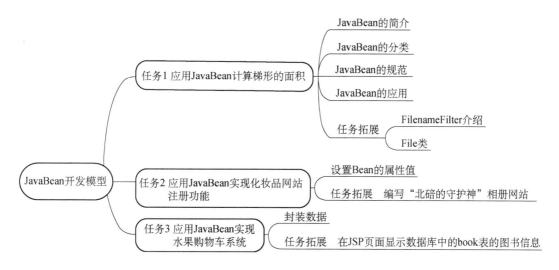

任务 1　应用 JavaBean 计算梯形的面积

任务演示

编写一个 JSP 页面，该页面提供一个表单，用户可以通过表单设置梯形的上底、下底和高的值，并提交给 JSP 页面，该 JSP 页面将计算梯形面积的任务交给 PageBean 完成。JSP 页面使用 getProperty 动作标记显示 PageBean 中的数据，计算梯形面积的结果如图 5-1 所示。

> http://localhost:8080/项目五/Area.jsp
> 梯形的上底:20.0
> 梯形的下底:30.0
> 梯形的高:5.0
> 梯形的面积:125.0

图 5-1　计算梯形面积的结果

知识准备

1. JavaBean 的简介

JavaBean 是 Java 的可重复使用软件组件，实际上 JavaBean 是一种 Java 类。每个 JavaBean 都实现了一个特定的功能，经过合理组织不同功能的 JavaBean，可以快速生成一个全新的应用程序。将一个应用程序比作一个房间，那么 JavaBean 就好比房间中的家具。由于 JavaBean 是基于 Java 语言的，因此 JavaBean 不依赖平台，它具有以下特点。

① 可以实现代码的重复利用。
② 易编写、易维护、易使用。
③ 可以在任何安装了 Java 运行环境的平台上使用，且不需要重新编译。

2. JavaBean 的分类

JavaBean 起初的功能是将可以重复使用的代码进行打包，在传统的应用中被应用于可视化界面，如 Java 图形界面中的按钮、文本框和列表框等都被称为可视化的 JavaBean。随着技术的发展与项目需求的增加，JavaBean 的功能与应用范围也在不断扩展。目前，JavaBean 主要用于实现业务逻辑或封装业务对象，这些 JavaBean 并没有可视化界面，因此又被称为非可视化的 JavaBean。

3. JavaBean 的规范

孟子曰"不以规矩，不能成方圆"，同样地，在设计 JavaBean 时也需要遵循有关约定，具体如下。

① 使用 package 语句打包。
② 在类中声明的方法的访问属性必须都是公有的（即用 public 修饰）。
③ 类中需提供公有的、无参数的构造方法。
④ 属性必须声明为私有的（即用 private 修饰）。

⑤ 如果类的成员变量名是 xxx，则类中需提供如下两个方法（将变量名放入方法名中时，要求变量名首字母大写）：

getXxx()：用来获取属性 xxx。

setXxx()：用来修改属性 xxx。

以上两个方法的名称都以"get"或"set"为前缀，以首字母大写的成员变量名的字符序列为后缀。

⑥ 对于 boolean 类型的成员变量（即布尔类型的属性），其类提供的方法名与⑤中的类似，但允许使用"is"代替"get"和"set"。

教学视频

【例 5-1】创建一个 Student 的 Bean 类，代码如下。

```
1  package my.bean;
2  public class Student {
3      private String name;
4      private int age;
5      public Student() {

6      }
7      public String getName() {
8          return name;
9      }
10     public void setName(String name) {
11         this.name = name;
12     }
13     public int getAge() {
14         return age;
15     }
16     public void setAge(int age) {
17         this.age = age;
18     }
19 }
```

代码解析：第 1 行定义了名为"my.bean"的包，并将 Student 类添加到了该包中；第 3~4 行定义了两个属性，分别代表姓名和年龄；第 5~6 行定义了一个不带参数的 Student() 构造方法；第 7~18 行分别为两个属性提供了两个方法。

【脚下留心】

在编写 JavaBean 时，需要将其放入某个包中才能被 JSP 页面访问。

4. JavaBean 的应用

通过 JavaBean 技术和 JSP 技术的结合，不仅可以实现表现层和逻辑层的分离，还可以提高 JSP 程序的运行效率和代码重用性，并且可以实现并行开发。JSP 提供了<jsp:useBean>、<jsp:getProperty>和<jsp:setProperty>动作元素来实现对 JavaBean 的操作。

（1）<jsp:useBean>动作元素

<jsp:useBean>动作元素可以定义一个具有一定生存范围和唯一 id 的 JavaBean 实例。JSP

页面可通过指定 id 来识别 JavaBean，也可以通过 id.method 语句调用 JavaBean 中的方法。在执行过程中，<jsp:useBean>动作元素首先会尝试寻找已经存在的，且具有相同 id 和 scope 值的 JavaBean 实例，如果没有找到，就会自动创建一个新的实例，其基本格式如下。

```
<jsp:useBean id="beanname" class="package.class" scope="page|request|session|application"/>
```

或

```
<jsp:useBean id=" beanname " class=" package.class " scope=" page| request| session|application ">
</jsp:useBean>
```

<jsp:useBean>动作元素的基本属性如表 5-1 所示。

表 5-1   <jsp:useBean>动作元素的基本属性

序号	属性名	功能
1	id	JavaBean 的唯一标识代表了一个 JavaBean 实例，它具有特定的 scope 的存在范围。JSP 页面通过 id 识别 JavaBean
2	scope	代表 JavaBean 的生存时间，scope 可以是 page、request 等，默认是 page
3	class	代表 JavaBean 的 class 名，需要特别注意的是，字母大小写要完全一致

（2）<jsp:getProperty>动作元素

<jsp:getProperty>动作元素用于获取 Bean 对象的属性值。JavaBean 实例必须在<jsp:getProperty>动作元素前面定义，其基本格式如下：

```
<jsp:getProperty name="id" property="Bean 属性" />
```

或

```
<jsp:getProperty name="id" property="Bean 属性"/>
</jsp:getProperty>
```

<jsp:getProperty>动作元素的基本属性如表 5-2 所示。

表 5-2   <jsp:getProperty>动作元素的基本属性

序号	属性名	功能
1	name	指出要获取哪个 Bean 的属性值，取值与 useBean 的 id 值相同
2	property	指出要获取 Bean 的哪个属性值

（3）<jsp:setProperty>动作元素

<jsp:setProperty>动作元素可以设置 JavaBean 的属性，它有两种基本格式。

格式一：将 JavaBean 属性的值设置为一个字符串。

```
<jsp:setProperty name="beanname" property="propertyname" value="propertyvalue"/>
```

格式二：将 JavaBean 属性的值设置为一个表达式的值。

```
<jsp:setProperty name="id" property="bean 的属性" value= "<%=expression%>"/>
```

<jsp:setProperty>动作元素的基本属性如表 5-3 所示。

表 5-3 <jsp:setProperty>动作元素的基本属性

序号	属性名	功能
1	name	指出要获取哪个 Bean 的属性值，取值与 useBean 的 id 值相同
2	property	代表要设置值的属性名。①当有"property="*""时，程序会反复查找当前所有的 ServletRequest 参数，匹配 JavaBean 中相同名称的属性，并通过 JavaBean 的属性的 set 方法给这个属性赋值为 value；②如果 value 的属性值为空，则不会修改 JavaBean 的属性值
3	param	代表页面请求的参数名，<jsp:setProperty>动作元素不能同时使用 param 和 value
4	value	代表赋给 JavaBean 的属性 property 的具体值

【例 5-2】编写 person.jsp 页面访问例 5-1 的 Bean 类的属性，代码如下。

```
1 <%@ page language="java" contentType="text/html;charset=UTF-8"
2 pageEncoding="UTF-8"%>
3 <jsp:useBean id="lili" class="my.bean.Student"
4 scope="request"></jsp:useBean>
5 <!DOCTYPE html PUBLIC "-//W3C//DTD HTML 4.01 Transitional//EN" 6 "http://www.w3.org/TR/html4/loose.dtd">
6 <html>
7 <head>
8 <meta http-equiv="Content-Type" content="text/html;charset=UTF-8">
10 <title>Insert title here</title>
11 </head>
12 <body bgcolor="f0fff0"><font size=4><b>
13 <jsp:setProperty property="name" name="lili" value="丽丽"/>
14 <jsp:setProperty property="age" name="lili" value="<%=20 %>"/>
15 学生的姓名:<jsp:getProperty property="name" name="lili"/><br>
16 学生的年龄:<jsp:getProperty property="age" name="lili"/>
17 </font></b>
18 </body>
19 </html>
```

代码解析：第 3 行通过<jsp:useBean>动作元素实例化 JavaBean 对象，id 是"lili"；第 13～14 行通过<jsp:setProperty>动作元素设置 JavaBean 中的 name 和 age 的属性值；第 15～16 行通过<jsp:getProperty>动作元素获取 JavaBean 的属性值。

任务实施

第一步：Bean 的类文件以"class"形式创建，此处创建的类文件为 AreaBean.java。这里还需要把 Bean 的类文件放到 my.bean 包中。

```
package my.bean;
public class AreaBean {
private double hight,up,down,area;
public AreaBean()
{

}
```

```java
    public double getHight() {
        return hight;
    }
    public void setHight(double hight) {
        this.hight = hight;
    }
    public double getUp() {
        return up;
    }
    public void setUp(double up) {
        this.up = up;
    }
    public double getDown() {
        return down;
    }
    public void setDown(double down) {
        this.down = down;
    }
    public double getArea() {
        this.area=(up+down)*hight/2;
        return area;
    }
    public void setArea(double area) {
        this.area = area;
    }
}
```

第二步：创建 JSP 文件。

```jsp
<%@ page language="java" contentType="text/html;charset=UTF-8"
    pageEncoding="UTF-8"%>
<jsp:useBean id="tixing" class="my.bean.AreaBean" scope="request"></jsp:useBean>
<!DOCTYPE html PUBLIC "-//W3C//DTD HTML 4.01 Transitional//EN" "http://www.w3.org/TR/html4/loose.dtd">
<html>
<head>
<meta http-equiv="Content-Type" content="text/html;charset=UTF-8">
<title>Insert title here</title>
</head>
<body>
<body bgcolor="f0fff0"><font size=4><b>
<jsp:setProperty property="up" name="tixing" value="20"/>
<jsp:setProperty property="down" name="tixing" value="30"/>
<jsp:setProperty property="hight" name="tixing" value="5"/>
梯形的上底:<jsp:getProperty property="up" name="tixing"/><br>
梯形的下底:<jsp:getProperty property="down" name="tixing"/><br>
梯形的高:<jsp:getProperty property="hight" name="tixing"/><br>
梯形的面积:<jsp:getProperty property="area" name="tixing"/><br>
</font></b>
</body>
```

```
</body>
</html>
```

任务拓展

通过本项目任务 1 的学习，我们已经学会使用简单的 Bean 了，在编写一个 Bean 的时候，除了需要应用 import 语句引入系统提供的类，可能还需要一些自己编写的类，只要将这些类和创建的 Bean 的类写在一个 Java 源代码中即可。

1. FilenameFilter 介绍

FilenameFilter 是一个接口，实现 FilenameFilter 接口的类的实例可用于过滤不符合规格的文件，并返回合格的文件，只有 accept()方法可用来测试指定的文件是否应包含在文件列表中，其基本格式如下。

```
public boolean accept(File dir,String name) {}
```

参数说明：
dir：表示文件的当前目录。
name：表示文件名。

2. File 类

File 类的一个对象代表一个文件或一个文件目录（即文件夹），基本格式如下：

```
File f=new File("路径");
```

常用的方法有以下两种。
① String[] fs = f.list()：用于获取指定目录下的所有文件或者文件目录的名称数组。
② File[] fs = f.listFiles()：用于获取指定目录下的所有文件或者文件目录的 File 数组。
FilenameFilter 用来返回符合要求的文件或目录，因此可以调用下面两种方法。
① String []fs = f.list（FilenameFilter filter）。
② File[]fs = f.listFiles（FilenameFilter filter）。
当向 list()方法传递一个 FilenameFilter 接口对象时，list()方法在列出文件时，会让该文件调用 accept()方法检查该文件是否符合 accept()方法指定的目录和文件的要求。

【例 5-3】应用 Bean 列出在 JSP 页面所在目录中具有特定扩展名的文件代码如下，运行结果如图 5-2 所示。

第一步：编写 FileName 类和 ListFile 类。

```
package my.bean;
import java.io.File;
import java.io.FilenameFilter;
  class FileName implements FilenameFilter{
  String str=null;
  FileName(String s)
  {
       str="."+s;
  }
  public boolean accept(File dir,String name) {
```

```java
        return name.endsWith(str);
    }
}
public class ListFile
{
    String extendsName=null;
    StringBuffer allFileName=new StringBuffer();
    public String getExtendsName() {
        return extendsName;
    }
    public void setExtendsName(String extendsName) {
        this.extendsName = extendsName;
    }
    public StringBuffer getAllFileName() {
        File dir=new File("D:/douban");
        FileName filename=new FileName(extendsName);
        String file_name[]=dir.list(filename);
        for(int i=0;i<file_name.length;i++)
        {
            allFileName.append(file_name[i]+" ");
        }
        return allFileName;
    }
}
```

第二步：编写 JSP 文件。

```jsp
<%@ page language="java" contentType="text/html;charset=UTF-8"
    pageEncoding="UTF-8"%>
<%@ page import="my.bean.ListFile" %>
<jsp:useBean id="file" class="my.bean.ListFile" scope="page"></jsp:useBean>
<!DOCTYPE html PUBLIC "-//W3C//DTD HTML 4.01 Transitional//EN" "http://www.w3.org/TR/html4/loose.dtd">
<html>
<head>
<meta http-equiv="Content-Type" content="text/html;charset=UTF-8">
<title>Insert title here</title>
</head>
<body bgcolor="pink">
<form action="" method="post">
```

第三步：输入文件的扩展名。

```jsp
<input type="text" name="extendsName">
<input type="submit" value="提交" >
</form>
<jsp:setProperty property="extendsName" name="file" param="extendsName"/>
<p>扩展名是<jsp:getProperty property="extendsName" name="file"/>的文件有：
<br><jsp:getProperty property="allFileName" name="file"/>
</body>
</html>
```

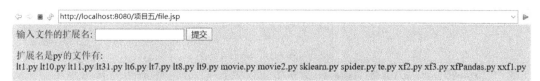

图 5-2 列出具有特定扩展名的文件的运行结果

任务 2　应用 JavaBean 实现化妆品网站注册功能

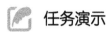 任务演示

当我们想在化妆品网站购买化妆品时,需要先注册才能购买,主界面、用户网站注册界面、显示用户网站注册信息界面分别如图 5-3、图 5-4、图 5-5 所示。

教学视频

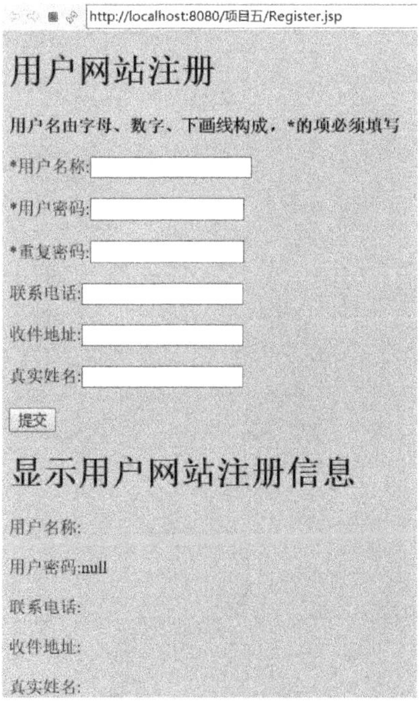

图 5-3　主界面

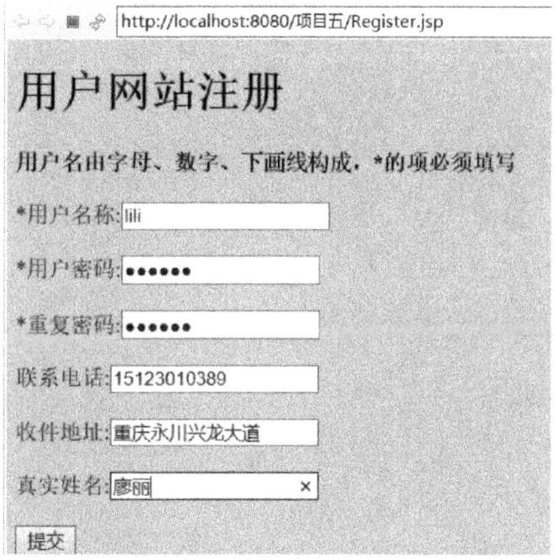

图 5-4　用户网站注册界面

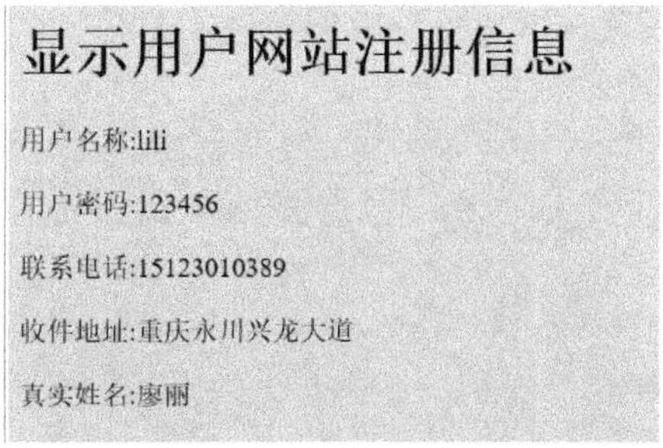

图 5-5　显示用户网站注册信息界面

知识准备

设置 Bean 的属性值

使用 setProperty 元素设置 Bean 的属性值可以通过 HTML 表单的参数值来实现，要求 Bean 属性的名称必须在表单中有相应的参数名与之对应，JSP 引擎会自动将参数的字符串类型值转换为与 Bean 对应的属性的值。

如果用 HTTP 表单的所有参数值来设置与 Bean 对应的属性值，则可以使用 setProperty 元素实现。具体包括以下三种方式。

方式一：<jsp:setProperty name="beanname" property="*" />。

setProperty 元素不再具体指定 Bean 属性的值将对应表单中的哪些参数指定的值，系统将自动根据名称进行匹配。

【例 5-4】通过 Bean 实现使用表单读取网站登录信息，代码如下，用户登录前后界面的对比如图 5-6 所示。

教学视频

第一步：编写 JavaBean 的 Login 类。

```
package my.bean;
public class Login {
private String username;
private String password;
public Login()
{
}
public String getUsername() {
    return username;
}
public void setUsername(String username) {
    this.username = username;
}
public String getPassword() {
    return password;
}
public void setPassword(String password) {
    this.password = password;
}
}
```

第二步：编写 JSP 页面。

```
1.  <%@ page language="java" contentType="text/html;charset=UTF-8"
2.     pageEncoding="UTF-8"%>
3.  <jsp:useBean id="lili" class="my.bean.Login" ></jsp:useBean>
4.  <!DOCTYPE html PUBLIC "-//W3C//DTD HTML 4.01 Transitional//EN"
5.  "http://www.w3.org/TR/html4/loose.dtd">
6.  <html>
7.  <head>
8.  <meta http-equiv="Content-Type" content="text/html;charset=UTF-8">
9.  <title>Insert title here</title>
10. </head>
```

```
11. <body bgcolor="pink">
12. <form action="" method="post">
13. 用户的登录界面<br>
14. 用户名:<input type="text" name="username"><br>
15. 密    码:<input type="password" name="password"><br>
16. <input type="submit" value="提交给本页面"><br>
17. </form>
18. 显示用户的登录信息<br>
19. <% request.setCharacterEncoding("utf-8");%>
20. <jsp:setProperty property="*" name="lili"/>
21. 用户名:<jsp:getProperty property="username" name="lili"/> <br>
22. 密码:<jsp:getProperty property="password" name="lili"/>
23. </body>
24. </html>
```

代码解析：第 3 行中的"<jsp:useBean>"定义了一个 id 为"lili"的 Register 实例；第 20 行中的"property="*""实现了对 HTML 表单元素与 Register.java 中的属性赋值，这种方法适合 HTML 表单中的元素的 name 属性值与 JavaBean 中的属性的属性名一致的情况。

图 5-6　用户登录前后界面的对比

方式二：<jsp:setProperty name="beanname" property="propertyName",param="parameterName"/>。

这种方式适合 HTML 表单中的元素的 name 属性值与 JavaBean 中的属性的属性名不一致的情况。如果将例 5-3 中的用户名的属性值设置为"user"，密码的属性值设置为"pwd"，则用方式一会出错，应用方式二不会出错，参考代码如下。

```
<jsp:setProperty property="username" name="lili" param="user"/>
<jsp:setProperty property="password" name="lili" param="pwd"/><br/>
```

上述代码中的 property 是 JavaBean 的属性，param 是表单控件的名称。

方式三：<jsp:setProperty name="beanname" property="propertyName",value="propertyValue"/>。

这种方式适合直接给指定的属性赋值，参考代码如下。

```
<jsp:setProperty property="username" name="lili" value="丽丽"/>
<jsp:setProperty property="password" name="lili" value="123456"/><br/>
```

任务实施

第一步：编写 JavaBean 的 Register 类。

```java
package my.bean;
public class Register {
    String logname="",password,phone="",address="",realname="";
public Register()
{

}
public String getLogname() {
    return logname;
}
public void setLogname(String logname) {
    this.logname = logname;
}
public String getPassword() {
    return password;
}
public void setPassword(String password) {
    this.password = password;
}
public String getPhone() {
    return phone;
}
public void setPhone(String phone) {
    this.phone = phone;
}
public String getAddress() {
    return address;
}
public void setAddress(String address) {
    this.address = address;
}
public String getRealname() {
    return realname;
}
public void setRealname(String realname) {
    this.realname = realname;
}
}
```

第二步：编写 JSP 页面。

```
<%@ page language="java" contentType="text/html;charset=UTF-8"
    pageEncoding="UTF-8"%>
 <jsp:useBean id="register" class="my.bean.Register" scope="request"> </jsp: useBean>
 <!DOCTYPE html PUBLIC "-//W3C//DTD HTML 4.01 Transitional//EN" "http://www.w3.org/TR/html4/loose.dtd">
 <html>
```

```html
<head>
<meta http-equiv="Content-Type" content="text/html;charset=UTF-8">
<title>Insert title here</title>
</head>
<body bgcolor="pink">
<div>
<h1>用户网站注册</h1>
    <form action="" method="post" name="form">
        <b> 用户名由字母、数字、下画线构成,带*的项必须填写 </b> <br><br>
        *用户名称:<input type="text" name="logname">  <br><br>
        *用户密码:<input type="password" name="password"> <br><br>
        *重复密码:<input type="password" name="again_password"> <br><br>
        联系电话:<input type="text" name="phone"> <br><br>
        收件地址:<input type="text" name="address"> <br><br>
        真实姓名:<input type="text" name="realname"> <br><br>
        <input type="submit" value="提交">
    </form></div>
<div>
<h1>显示用户网站注册信息</h1>
<%request.setCharacterEncoding("utf-8");%>
<jsp:setProperty property="*" name="register"/>
用户名称:<jsp:getProperty property="logname" name="register"/><br><br>
用户密码:<jsp:getProperty property="password" name="register"/><br><br>
联系电话:<jsp:getProperty property="phone" name="register"/><br><br>
收件地址:<jsp:getProperty property="address" name="register"/><br><br>
真实姓名:<jsp:getProperty property="realname" name="register"/><br><br>
</div></body>
</html>
```

任务拓展

编写"北碚的守护神"相册网站

任务要求:2022年8月,重庆市北碚区歇马街道虎头村突发山火,从山火烧起来的那一刻,各方力量开始集结,消防救援人员和志愿者们为了人民的生命安全与财产安全,在极端高温天气下处于炙烤险峻的一线,他们勇往直前、迎难而上,他们是最美的逆行者。请应用所学的知识,编写一个"北碚的守护神"相册网站,首页如图5-7所示。

图5-7 "北碚的守护神"相册网站首页

第一步：创建 FileName1 类和 Imagebean 类。

```java
package my.bean;
import java.io.*;
class FileName1 implements FilenameFilter
{
    public boolean accept(File dir,String name) {
        boolean boo=false;
        if(name.endsWith(".jpg")||name.endsWith(".JPG"))
            boo=true;
        return boo;
    }
}
public class Imagebean {
int imageNumber=0,max;
String pictureName[]={},playImage;
    public Imagebean()
    {
        File dir=new File("D:/workspace/项目五/WebContent/image");
        pictureName=dir.list(new FileName1());
        max=pictureName.length;
    }
    public int getImageNumber() {
        return imageNumber;
    }
    public void setImageNumber(int n) {
        if(n<0)
            n=max-1;
        if(n==max)
            n=0;
        imageNumber=n;
    }
    public String getPlayImage() {
        playImage=new String("<img src=image/"+pictureName[imageNumber]+" "+" width=500 height=300></img>");
        return playImage;
    }
}
```

第二步：创建 JSP 页面。

```jsp
<%@ page language="java" contentType="text/html;charset=UTF-8"
    pageEncoding="UTF-8"%>
<%@ page import="my.bean.*" %>
<jsp:useBean id="photo" class="my.bean.Imagebean" scope="session"></jsp:useBean>
```

```html
<!DOCTYPE html PUBLIC "-//W3C//DTD HTML 4.01 Transitional//EN" "http://www.w3.org/TR/html4/loose.dtd">
<html>
<head>
<meta http-equiv="Content-Type" content="text/html;charset=UTF-8">
<title>Insert title here</title>
</head>
<body bgcolor="pink">
<jsp:setProperty property="imageNumber" name="photo" param="imageNumber"/>
<h1>北碚的守护神</h1>
<table>
<form action="" method="post">
<tr>
<td><input type=submit name="up" value="上一张"></td>
<input type="hidden" name="imageNumber" value="<%=photo.getImageNumber()-1%>">
</form>
<form action="" method=post>
<td><input type=submit name="down" value="下一张"></td>
<input type="hidden" name="imageNumber" value="<%=photo.getImageNumber()+1%>">
</tr>
</form>
</table>
<jsp:getProperty property="playImage" name="photo"/>
</body>
</html>
```

任务 3　应用 JavaBean 实现水果购物车系统

任务演示

为了进一步提高水果销售业绩，便利客户购买商品，请应用 JavaBean 技术实现水果购物车系统，该系统的主要功能包含显示水果名称、购买水果、移除水果、清空购物车等，水果列表界面和购买商品界面分别如图 5-8、图 5-9 所示。

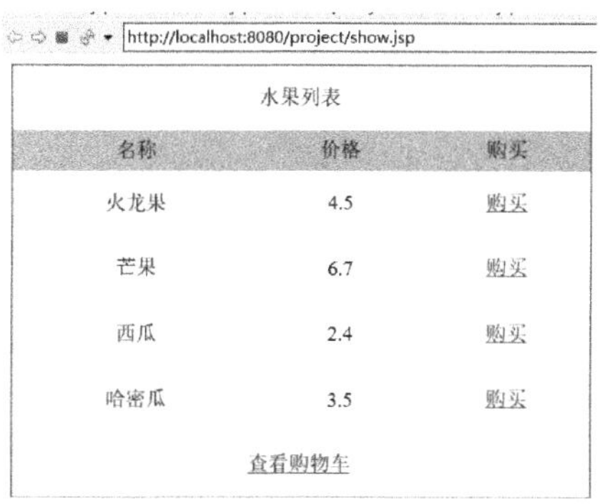

图 5-8　水果列表界面

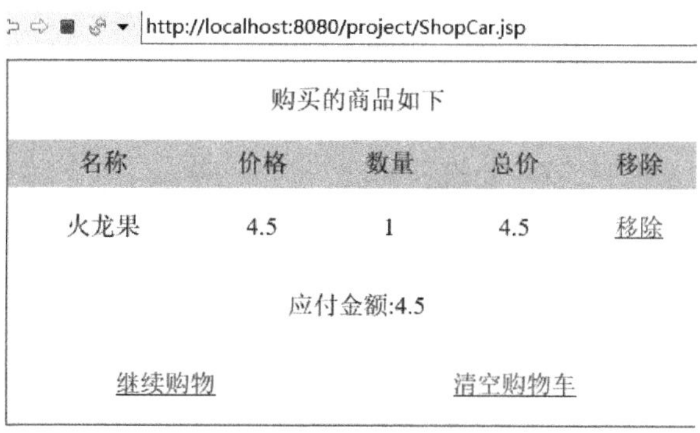

图 5-9　购买商品界面

知识准备

封装数据

适用于封装业务的 JavaBean 是完成一定运算和操作功能的业务类。JavaBean 在一定程度上可以将 Java 代码从 JSP 页面中分离。封装数据的 JavaBean 要将表单中的用户输入值传入数据库中的相应字段，或将数据库中的字段值取出并显示到网页中，这需要一个专用的 Bean 与封装数据的 JavaBean 配合完成。

项目四已经详细介绍了在 JSP 中连接数据库的多种方法，以及如何对数据库进行增加、删除、修改和查询等操作。在同一个应用程序中，许多地方都需要进行数据库连接和数据库内容更新，此时可以通过所学的 JavaBean 技术将数据库的一些操作封装到 Bean 中，当需要用到这些功能再直接用 JavaBean 的动作元素来完成 Bean 的调用，具体的知识点在前面已经介绍了，这里不再赘述。

 任务实施

第一步：创建封装商品信息的 JavaBean。

```java
package my.bean;
public class Goods {
private String name; //商品名称
private float price;//商品价格
private int num; //购买商品的数量
public Goods() {
    super();
}
public Goods(String name) {
    super();
    this.name = name;
}
public Goods(String name,float price,int num) {
    super();
    this.name = name;
    this.price = price;
    this.num = num;
}
public String getName() {
    return name;
}
public void setName(String name) {
    this.name = name;
}
public float getPrice() {
    return price;
}
public void setPrice(float price) {
    this.price = price;
```

```
}
public int getNum() {
    return num;
}
public void setNum(int num) {
    this.num = num;
}}
```

第二步：创建 JavaBean 工具，用于完成对中文乱码的处理和类型转换。

```
package my.bean;
public class MyTools {
//将字符串型转换为整型
public static int strToint(String str)
{
    if(str==null||str.equals(""))
        str="0";
    int num=0;
    try {
        num=Integer.parseInt(str);
    } catch (Exception e) {
        num=0;
        e.printStackTrace();
    }
    return num;
}
//处理中文乱码
public static String toChinese(String str)
{
    if(str==null)
        str="";
    try {
        str=new String(str.getBytes("ISO-8859-1"),"utf-8");
    } catch (Exception e) {
        str="";
        e.printStackTrace();
    }
    return str;
}
}
```

第三步：创建购物车，主要功能是购买水果、移除水果、清空购物车。

```
package my.bean;
import java.util.*;
import my.bean.*;
public class ShopCar {
private ArrayList buylist=new ArrayList();
public ArrayList getBuylist() {
```

```java
        return buylist;
}
public void setBuylist(Goods list) {  //购买水果
    if(list!=null)
    {
        if(buylist.size()==0)
        {
            Goods temp=new Goods();
            temp.setName(list.getName());
            temp.setPrice(list.getPrice());
            temp.setNum(list.getNum());
            buylist.add(temp);
        }else
        {
            int i=0;
            for(;i<buylist.size();i++)
            {
                Goods temp=(Goods)buylist.get(i);
                if(temp.getName().equals(list.getName()))
                {
                    temp.setNum(temp.getNum()+1);
                    break;
                }
            }
            if(i>buylist.size())
            {
                Goods temp=new Goods();
                temp.setName(list.getName());
                temp.setPrice(list.getPrice());
                temp.setNum(list.getNum());
                buylist.add(temp);
            }
        }
    }
}
public void removeItem(String name)  //移除水果
{
    for(int i=0;i<buylist.size();i++)
    {
        Goods temp=(Goods)buylist.get(i);
        if(temp.getName().equals(MyTools.toChinese(name)))
        {
            if(temp.getNum()>1)
            {
                temp.setNum(temp.getNum()-1);
                break;
            }
```

```
                else if(temp.getNum()==1)
                {
                    buylist.remove(i);
                }
            }
        }}
    public void clearCar()  //清空购物车
    {
        buylist.clear();
    }
}
```

第四步：创建水果列表页面。

```jsp
<%@ page language="java" contentType="text/html;charset=UTF-8"
    pageEncoding="UTF-8"%>
    <%@ page import="my.bean.Goods" %>
    <%@ page import="java.util.*" %>
<!DOCTYPE html PUBLIC "-//W3C//DTD HTML 4.01 Transitional//EN" "http://www.w3.org/TR/html4/loose.dtd">
<html>
<head>
<meta http-equiv="Content-Type" content="text/html;charset=UTF-8">
<title>Insert title here</title>
</head>
<body>
<%!
static ArrayList goodslist=new ArrayList();
static{
    String[] names={"火龙果","芒果","西瓜","哈密瓜"};
    float[] prices={4.5f,6.7f,2.4f,3.5f};

for(int i=0;i<4;i++){
    Goods list1=new Goods();
    list1.setName(names[i]);
    list1.setPrice(prices[i]);
    list1.setNum(1);
    goodslist.add(i,list1);
}
}
%>
<%
session.setAttribute("goodslist",goodslist);
response.sendRedirect("show.jsp");
%>
</body>
</html>
```

第五步:展示水果列表界面。

```jsp
<%@ page language="java" contentType="text/html;charset=UTF-8"
    pageEncoding="UTF-8"%>
<%@ page import="java.util.ArrayList" %>
<%@ page import="my.bean.Goods" %>

<!DOCTYPE html PUBLIC "-//W3C//DTD HTML 4.01 Transitional//EN" "http://www.w3.org/TR/html4/loose.dtd">
<html>
<head>
<meta http-equiv="Content-Type" content="text/html;charset=UTF-8">
<title>Insert title here</title>
</head>
<body>
<% ArrayList goodslist=(ArrayList)session.getAttribute("goodslist");%>
 <table border="1" width="450" rules="none" cellpadding="0" cellspacing="0">
 <tr height="50">
 <td colspan="3" align="center">水果列表</td>
  </tr>

 <tr align="center" height="30" bgcolor="lightgrey">
  <td>名称</td>
  <td>价格</td>
  <td>购买</td>
  </tr>
  <% if(goodslist==null||goodslist.size()==0){ %>
 <tr><td>没有水果了!</td></tr>
  <%} else { for(int i=0;i<goodslist.size();i++){
  Goods list=(Goods)goodslist.get(i);
  %>
  <tr height="50" align="center">
   <td><%=list.getName() %></td>
   <td><%=list.getPrice() %></td>
   <td><a href="docar.jsp?action=buy&id=<%=i%>">购买</a></td>
  </tr>
  <%}} %>
  <tr height="50">
   <td align="center" colspan="3"><a href="ShopCar.jsp">查看购物车</a></td>
  </tr>
 </table>
</body>
</html>
```

第六步:编写水果购物车系统页面。

```jsp
<%@ page language="java" contentType="text/html;charset=UTF-8"
    pageEncoding="UTF-8"%>
```

```jsp
        <%@ page import="java.util.ArrayList" %>
<%@ page import="my.bean.*" %>
<jsp:useBean id="myCar" class="my.bean.ShopCar" scope="session"></jsp:useBean>
<!DOCTYPE html PUBLIC "-//W3C//DTD HTML 4.01 Transitional//EN" "http://www.w3.org/TR/html4/loose.dtd">
<html>
<head>
<meta http-equiv="Content-Type" content="text/html;charset=UTF-8">
<title>Insert title here</title>
</head>
<body>
<%
String action=request.getParameter("action");
if(action==null)
    action="";
if(action.equals("buy"))
{
    ArrayList goodslist=(ArrayList)session.getAttribute("goodslist");
    int id=MyTools.strToint(request.getParameter("id"));
    Goods list=(Goods)goodslist.get(id);
    myCar.setBuylist(list);
    response.sendRedirect("show.jsp");
}
else if(action.equals("remove"))
{
    String name=request.getParameter("name");
    myCar.removeItem(name);
    response.sendRedirect("ShopCar.jsp");
}else if(action.equals("clear"))
{
    myCar.clearCar();
    response.sendRedirect("ShopCar.jsp");
}
else
{
    response.sendRedirect("show.jsp");
}
%>
</body>
</html>
```

任务拓展

下面在JSP页面显示数据库中的book表的图书信息。

任务要求：通过数据库获取数据，并显示在 JSP 页面中，显示结果如图 5-10 所示。

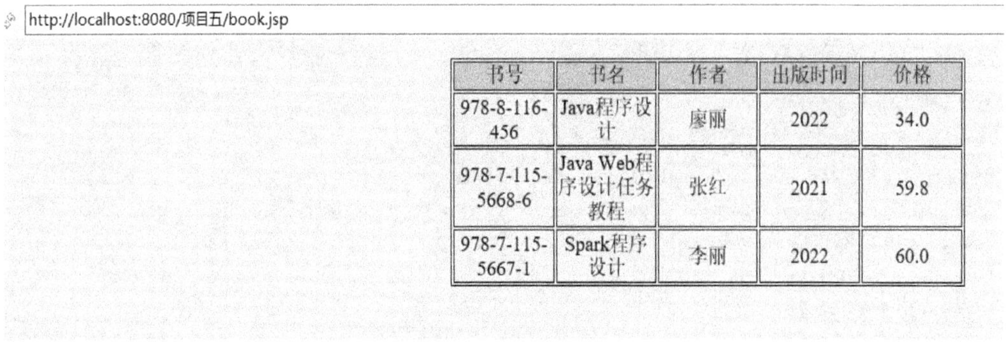

书号	书名	作者	出版时间	价格
978-8-116-456	Java程序设计	廖丽	2022	34.0
978-7-115-5668-6	Java Web程序设计任务教程	张红	2021	59.8
978-7-115-5667-1	Spark程序设计	李丽	2022	60.0

图 5-10 显示结果

第一步：在 MySQL 数据库中创建 book 数据库，并在该数据库中创建如表 5-4 所示的 book 表。

表 5-4 book 表

字段（属性）	类型	说明
isbn	varchar(20)	书号
bookname	varchar(20)	书名
author	varchar(20)	作者
publicdate	int	出版时间
price	double	价格

第二步：创建一个 BookBean 类，要求包含 book 表的 5 个属性和相应的 set 方法、get 方法。

```java
package my.bean;
import java.sql.Connection;
import java.sql.DriverManager;
import java.sql.SQLException;

public class BookBean {
private String isbn;
private String bookname;
private String author;
private int publicdate;
private double price;
Connection conn = null;
public String getIsbn() {
    return isbn;
}
public void setIsbn(String isbn) {
    this.isbn = isbn;
}
public String getBookname() {
```

```java
        return bookname;
    }
    public void setBookname(String bookname) {
        this.bookname = bookname;
    }
    public String getAuthor() {
        return author;
    }
    public void setAuthor(String author) {
        this.author = author;
    }

    public int getPublicdate() {
        return publicdate;
    }
    public void setPublicdate(int publicdate) {
        this.publicdate = publicdate;
    }
    public double getPrice() {
        return price;
    }
    public void setPrice(double price) {
        this.price = price;
    }
    public Connection dbconnection() {
        String dburl = "jdbc:mysql://127.0.0.1:3306/books?useSSL=false&serverTimezone=UTC";
        String username ="root";
        String password = "root";
        try{
            Class.forName("com.mysql.cj.jdbc.Driver");
            conn = DriverManager.getConnection(dburl,username,password);
            System.out.println("数据库连接成功");
             return conn;

        }catch (ClassNotFoundException e1){
            System.out.println(e1+"驱动程序找不到");
        }catch(SQLException e2){
            System.out.println(e2);
        }
        return null;
    }
}
```

第三步：实现数据库连接，并显示数据库的数据。

```jsp
<%@ page language="java" contentType="text/html;charset=UTF-8"
    pageEncoding="UTF-8"%>
<%@ page import="java.sql.*" %>
<%@ page import="my.bean.*" %>
```

```jsp
<jsp:useBean id="dbcon" class="my.bean.DBcon" scope="request"></jsp:useBean>
<!DOCTYPE html PUBLIC "-//W3C//DTD HTML 4.01 Transitional//EN" "http://www.w3.org/TR/html4/loose.dtd">
<html>
<head>
<meta http-equiv="Content-Type" content="text/html;charset=UTF-8">
<title>图书信息</title>
<style>
    body{
        background-color:rgba(0,255,255,0.205);
    }
    table{
        margin:20px auto;
    }
    td{
        width:100px;
        text-align:center;
    }
        .aa{
        background-color:rgba(0,0,0,0.178);
    }
    .bb{
        background-color:rgba(235,145,228,0.178);
    }
</style>

</head>
<body>
<table border="1">
<tr class="aa">
<td>书号</td>
<td>书名</td>
<td>作者</td>
<td>出版时间</td>
<td>价格</td>
</tr>
<%
BookBean aa = new BookBean();
Connection db=aa.dbconnection();
Statement stmt=db.createStatement();
ResultSet rs=stmt.executeQuery("select * from book");
while(rs.next())
{
%>
<tr class="bb">
<td><%=rs.getString(1) %></td>
<td><%=rs.getString(2) %></td>
<td><%=rs.getString(3) %></td>
```

```
<td><%=rs.getInt("publicdate") %></td>
<td><%=rs.getDouble("price") %></td>
</tr>
<%}
rs.close();
stmt.close();
db.close();
%>
</table>
</body>
</html>
```

项 目 实 训

实训一　应用 JavaBean 技术设计留言板

要求：通过表单提交留言信息，并显示留言信息，留言板界面、留言板信息显示效果分别如图 5-11、图 5-12 所示。

图 5-11　留言板界面

图 5-12　留言板信息显示效果

实训二　应用 JavaBean 技术实现四则运算

要求：在文本框内输入数字，并选择运算的类型，将输入的数字及运算结果显示在界面上，四则运算界面如图 5-13 所示。

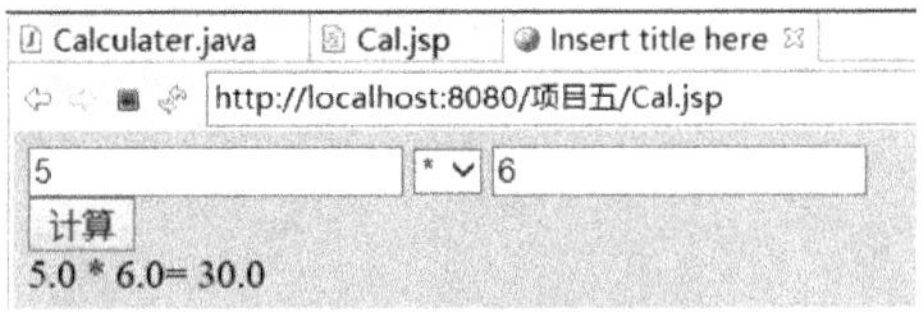

图 5-13 四则运算界面

实训三　为化妆品网站设计查看功能

要求：先连接数据库，再从数据库中获取化妆品名，最后显示在界面上，查看化妆品的界面如图 5-14 所示。

图 5-14 查看化妆品的界面

课 后 练 习

一、填空题

1. _____技术和 JSP 技术相结合，可以实现表现层和逻辑层的分离。
2. 在 JSP 中可以使用_____操作来设置 Bean 的属性，也可以使用_____操作来获取 Bean 的值。
3. JavaBean 有四个 scope，它们分别是_____、_____、_____和_____。
4. _____动作元素被用来创建一个 Bean 实例，并指定它的名称和作用范围。
5. _____动作元素用来设置 Bean 的属性值。
6. _____动作元素用来获得 Bean 的属性值。

二、选择题

1. 在"<jsp:useBean id="bean 的名称",scope="bean 的有效范围"class="包名.类名"/>动作标记"中，scope 的值不可以是（　　）。
   A. page　　　　　　B. request　　　　　　C. session　　　　　　D.response
2. 关于 JavaBean 的说法正确的是（　　）。
   A. Java 文件与 Bean 定义的类可以不同名，但一定要注意区分字母大小写
   B. 在 JSP 文件中引用 Bean，其实就是用<jsp:useBean>动作元素
   C. 被引用的 Bean 文件的文件名后缀为".java"
   D. Bean 文件放在任何目录下都可以被引用
3. 在项目中已经建立了一个 JavaBean，该类为 bean.Student，其中的 bean 具有 name 属性，下面动作元素的用法正确的是（　　）。

A. <jsp:useBean id="student" class="Student" scope="session"></jsp:useBean>
B. <jsp:useBean id="student" class="Student" scope="session">helo student!</jsp:useBean>
C. <jsp:useBean id="student" class="bean.Student" scope="session">hello student!</jsp:useBean>
D. <jsp:getProperty class="name"property="student"></jsp:useBean>

4. (　　) 能使 Bean 一直保留至到期或被删除。
A. page　　　　　B. session　　　　　C. application　　　　　D. request

5. tom.jiafei.Circle 是创建 Bean 的类，下列哪个是正确创建 session 周期的 Bean 的选项？(　　)
A. <jsp:useBean id="circle" class="tom.jiafei.Circle" scope="page"/>
B. <jsp:useBean id="circle" class="tom.jiafei.Circle" scope="request"/>
C. <jsp:useBean id="circle" class="tom.jiafei.Circle" scope="session"/>
D. <jsp:useBean id="circle" class="tom.jiafei.Circle" scope="application"/>

6. 假设创建 Bean 的类有一个 int 类型的属性 number，下列哪个方法是设置该属性值的正确方法？(　　)
A. public void setNumber(int n)
　　{ number=n；}
B. void setNumber(int n)
　　{ number=n；}
C. pubic void SetNumber(int n)
　　{ number=n}
D. public void Setnumber(int n)
　　{ number=n；}

8. (　　) 方法可用于获取 Bean 的属性值。
A. segProperty　　　B. setValue　　　C. getProperty　　　D. getValue

9. 在 JSP 中调用 JavaBean 时不会用到的动作元素是 (　　)。
A. <javabean>　　　　　　　　　　B. <jsp:useBean>
C. <jsp:setProperty>　　　　　　　D. <jsp:getProperty>

10. (　　) 是一种可以在一个或多个应用程序中重复使用的组件。
A. JSP 页面　　　B. JavaMail　　　C. JavaBean　　　D. Servlet

11. JavaBean 的属性可以使用 (　　) 来访问。
A. 属性　　　　　　　　　　　　B. get 方法和 set 方法
C. 事件　　　　　　　　　　　　D. Scriptlet

三、编程题

1. 实现一个简单的登录程序。要求应用 JavaBean 接收用户输入的用户名和密码，判断输入的用户名是否为"admin"、密码是否为"123"，若都是，则转发到 success.jsp 页面，并显示"欢迎登录"的提示信息，否则转发到 fault.jsp 页面，并显示"登录失败"的提示信息。

2. 编写一个 car.jsp 汽车信息页面，从表单中输入汽车的牌号、名称和生产日期，应用 JavaBean 实现汽车信息的显示。

项目六　Servlet 技术

项目要求

本项目以 Tomcat 为例介绍 Java Web 技术如何基于 Servlet 进行工作，并进一步利用 Servlet 规范的三大高级特性（Filter、Listener、文件上传与下载）实现对 request 对象、response 对象的拦截和修改，完成对 context、session、request 事件的监听和处理，并结合 Commons-FileUpload 组件实现文件上传与下载的功能。

项目分析

要完成项目任务，至少需要具备两个基本条件：一是掌握 Servlet 基本知识及其工作原理，二是掌握 Servlet 规范的三大高级特性。本项目分为 3 个任务：创建并配置 Servlet 程序、应用 HttpServletRequest 对象和 HttpServletResponse 对象实现用户验证登录、应用 Commons-FileUpload 和 Filter 实现文件上传与下载。

项目目标

【知识目标】掌握 Servlet 的基本概念、特点和接口；掌握 Servlet 的配置和生命周期；掌握 ServletConfig 对象和 ServletContext 对象的使用方法；掌握 HttpServletRequest 对象和 HttpServletResponse 对象的使用方法；掌握 Filter 的映射与过滤器，以及 Listener 的使用方法；掌握文件上传和下载的流程。

【能力目标】能熟练使用 Eclipse 工具开发 Servlet、Filter、Listener 程序；能够熟练使用 Filter 对 request 对象、response 对象进行拦截，实现特定功能；能够熟练使用 Commons-FileUpload 组件实现文件上传与下载；会使用 Listener 监听 ServletContext 对象、HttpSession 对象、ServletRequest 对象的相关事件。

【素质目标】培养学生的团队合作意识和自主学习能力。

知识导图

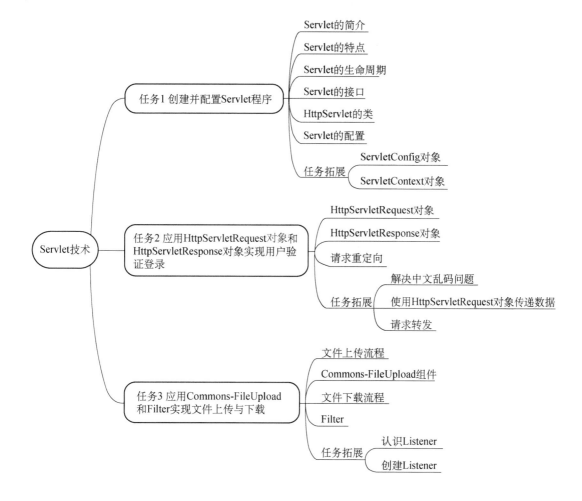

任务 1 创建并配置 Servlet 程序

任务演示

2020 年刊发的《自主创新，中国软件从追赶到超越》报道了金蝶国际软件集团有限公司（简称金蝶）敢为人先，在敢闯中"试"出了多项第一，创造了辉煌的历史。一路闯来的金蝶，今天依然没有停止创新、超越的脚步，因为金蝶深知"沿着旧地图，找不到新大陆"，唯有勇于自我颠覆，不断超越自我，才能引领行业的发展，帮助广大中国企业智慧成长，不负时代，助力中国梦早日实现。

在本次任务中，我们将创建一个 Servlet 程序，并对其进行配置，使 Servlet 程序能在服务器中正确运行，并能处理请求信息，最终动态生成一个网页，在该网页上显示"自主创新，中国软件从追赶到超越"，显示结果如图 6-1 所示。

图 6-1　显示结果

 知识准备

1. Servlet 的简介

Servlet 是运行在 Web 服务器上的 Java 应用程序，用于处理客户端的 Servlet 请求，具有动态生成网页、实现与其他服务器资源（如数据库）进行通信等的功能。Servlet 程序必须运行在 Servlet 容器中（如 Tomcat、Jetty、Resin 等），Servlet 容器负责按照规则加载、运行 Servlet 程序。用户访问 Servlet 程序的流程如图 6-2 所示。

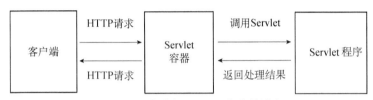

图 6-2　用户访问 Servlet 程序的流程

2. Servlet 的特点

Servlet 不仅具有 Java 语言的特点，还封装了与 Web 相关的应用，因此 Servlet 在功能、性能、安全性等方面都十分优秀。Servlet 的特点表现在以下五个方面。

① 功能强大：Servlet 不仅可以调用 Java API 中的对象及方法，还可以调用 Servlet 封装的 Servlet API 接口。因此，Servlet 的功能十分强大。

② 可移植：Servlet 使用 Java 语言编写，而 Java 语言是跨平台的，因此 Servlet 可以在多个系统平台中运行。

③ 性能高：Servlet 只在第一次被请求时进行实例化，后续便常驻内存中，以后的请求都由该实例进行处理，即每一个请求对应一个线程，而不是对应进程。

④ 安全性高：Servlet 不仅使用了 Java 的安全框架，还使用了 Servlet 容器提供的安全功能，因此安全性非常高。

⑤ 可扩展：由于 Java 语言是面向对象的编程语言，所以 Servlet 可以通过封装、继承等特性扩展实际的业务需要。

3. Servlet 的生命周期

Servlet 是使用 Java 语言编写的应用程序，Java 语言的所有对象都有生命周期。Servlet 的生命周期总体分为 3 个阶段，按照先后顺序排列，分别是初始化阶段（只执行 1 次）、运行阶段（重复执行）、销毁阶段（只执行 1 次）。

（1）初始化阶段

当客户端向 Servlet 容器发出 HTTP 请求，指明要访问某个 Servlet 对象时，Servlet 容器首先会解析 HTTP 请求，检查内存中是否已经有该 Servlet 对象，如果有则直接使用该

Servlet 对象；如果没有，就创建 Servlet 对象。然后通过调用 init()方法完成 Servlet 对象的初始化，如建立数据库连接、获取配置信息等。注意，在 Servlet 对象的整个生命周期内，init()方法只能被调用一次。此外，Servlet 对象一旦创建就会驻留在内存中，等待客户端的访问，直到服务器关闭或 Web 应用被移出容器时，Servlet 对象才会被销毁。

（2）运行阶段

在运行阶段，Servlet 容器会先将客户端的 HTTP 请求封装成 ServletRequest 对象，并创建一个 ServletResponse 对象，该 ServletResponse 对象用于表示经过 Servlet 处理后生成的响应结果。然后，Servlet 容器会将创建的 ServletRequest 对象和 ServletResponse 对象作为参数，传递给 Servlet 程序的 service()方法。由此，service()方法就可以从传递过来的 ServletRequest 对象中获取客户端的请求信息并进行处理，并通过传递过来的 ServletResponse 对象生成返回给客户端的响应结果。注意，在其生命周期内，service()方法会被反复调用，即对每一次 Servlet 的访问请求，Servlet 容器都会创建新的 ServletRequest 对象和 ServletResponse 对象作为参数，并调用、执行一次 Servlet 的 service()方法。

（3）销毁阶段

当 Servlet 对象所在的 Web 应用程序被移除，或者服务器被关闭时，Servlet 容器会调用 Servlet 对象的 destroy()方法，以便让 Servlet 对象释放它占用的资源。注意，在 Servlet 对象的整个生命周期中，destroy()方法也只能被调用一次。

4. Servlet 的接口

Sun Microsystems 公司为开发人员提供了 Servlet 开发相关的一系列接口和类，其中最重要的是 Javax.servlet.Servlet 接口，所有 Servlet 类必须实现这个接口，才能接收和响应用户的请求。在 Servlet 接口中定义了 5 个接口方法，用于初始化、运行、销毁 Servlet 对象，具体如表 6-1 所示。

表 6-1　Servlet 接口定义的方法

方法	功能描述
void init(ServletConfig config)	在 Web 服务器（Servlet 容器）创建好 Servlet 对象后，就会调用该方法初始化 Servlet 对象
void service(ServletRequest request, ServletResponse response)	负责响应用户的请求，当 Web 服务器（Servlet 容器）接收到客户端访问特定 Servlet 对象的请求时，就会调用该 Servlet 对象的 service()方法，即 service()方法是 Servlet 应用程序的入口，相当于 Java 程序的 main()方法。注意，Web 服务器传入 service()方法的参数有两个：ServletRequest（即 JSP 中的 request 对象）和 ServletResponse（即 JSP 中的 response 对象）。前者实现了 HTTPServletRequest 接口，封装了浏览器向服务器发送的请求；后者实现了 HTTPServletResponse 接口，封装了服务器向浏览器返回的信息
void destroy()	负责释放 Servlet 对象占用的资源，当关闭 Web 服务器或者移除 Servlet 对象时，则调用该方法
ServletConfig getServletConfig()	返回一个 ServletConfig 对象，该对象包含了 Servlet 对象的初始化参数信息
String getServletInfo()	返回一个字符串，该字符串包含了 Servlet 对象的创建者、版本和版权等信息

注意：表 6-1 中的 init()、service()、destroy()方法都由 Servlet 容器调用，而且 Servlet 容器会在其生命周期的不同阶段分别调用上述 3 个方法，因此我们可以通过上述 3 个方法表现 Servlet 容器的生命周期。

5. HttpServlet 的类

Sun Microsystems 公司专门为 Servlet 接口提供了两个默认的接口实现类：GenericServlet 和 HttpServlet。GenericServlet 是一个抽象类，该类为 Servlet 接口提供了部分实现，但它并没有实现对 HTTP 请求的处理。而 HttpServlet 类是 GenericServlet 类的子类，它继承了 GenericServlet 类的所有方法，并且为 HTTP 请求中的 POST、GET 等类型提供了具体的操作方法。因此，开发人员编写的 Servlet 类一般都继承自 HttpServlet 类，即在实际开发中使用的具体的 Servlet 对象就是 HttpServlet 对象。在使用 HttpServlet 类的实例对象处理客户端的 HTTP 请求时，HTTP 的不同请求类型会涉及到如表 6-2 所示的两个常用方法。

表 6-2 HttpServlet 类的常用方法

方法	功能描述
protected void doGet(HttpServletRequest req,HttpServletResponse resp)	用于处理 GET 类型的 HTTP 请求
protected void doPost(HttpServletRequest req,HttpServletResponse resp)	用于处理 POST 类型的 HTTP 请求

6. Servlet 的配置

对 Servlet 进行配置后，Servlet 才能正确运行并处理请求信息。Servlet 的配置可以通过使用配置文件 web.xml 和使用@WebServlet 注解两种方式实现。注意，如果在 web.xml 文件中配置多个 Servlet 会非常烦琐，因此建议使用 Servlet3.0 之后的版本提供的@WebServlet 注解来简化 Servlet 的配置。

在部署项目时，Servlet 容器会根据@WebServlet 注解的属性配置将相应的类部署为 Servlet，@WebServlet 注解的常用属性如表 6-3 所示。

表 6-3 @WebServlet 注解的常用属性

属性声明	功能
String name	name 属性用于指定该 Servlet 的名称，一般与 Servlet 类名相同，要求具有唯一性
String[] value	value 属性用于映射访问对应 Servlet 的 URL 地址，URL 地址前必须加 "/"，否则无法访问。注意：value 属性等价于 urlPatterns 属性，并且 urlPatterns 属性和 value 属性不能同时使用
String[] urlPatterns	urlPatterns 属性用于指定一组 Servlet 的 URL 匹配模式。注意：urlPatterns 属性等价于 value 属性，并且 urlPatterns 属性和 value 属性不能同时使用
int loadOnStartup	loadOnStartup 属性用于指定 Servlet 的加载顺序
WebInitParam[]	该属性用于指定一组 Servlet 初始化参数
boolean asyncSupported	asyncSupported 属性用于声明 Servlet 是否支持异步操作模式
String description	Servlet 的描述信息
String displayName	Servlet 的显示的名称

注意：任何一个继承了 HttpServlet 类的子类都可以使用@WebServlet 注解进行标注。

【例 6-1】配置 Servlet。

```
@WebServlet(name = "TestServlet",urlPatterns = "/TestServlet")
public class TestServlet extends HttpServlet{
    //处理 GET 请求的方法
    public void doGet(HttpServletRequest request,HttpServletResponse response)
```

```
throws ServletException,IOException {}
    //处理 POST 请求的方法
    protected void doPost(HttpServletRequest request,HttpServletResponse
        response) throws ServletException,IOException {}}
```

代码解析：TestServlet 类继承了 HttpServlet 类，我们在该类的上方使用@WebServlet 注解将其标注为一个 Servlet，name 属性指定了该 Servlet 的名称为"TestServlet"，urlPatterns 属性指定了访问该 Servlet 的 URL 地址。注意，如果需要在@WebServlet 注解中设置多个属性，属性之间要用逗号隔开，如该例中的 name 属性和 urlPatterns 属性之间使用逗号隔开。

 任务实施

任务要求：
① 使用 Eclipse 创建一个 Servlet 程序。
② 使用@WebServlet 注解对该 Servlet 程序进行配置。
③ 请求访问该 Servlet 程序，并动态生成一个网页，在该网页上显示"自主创新，中国软件从追赶到超越"。

教学视频

第一步：在"New Dynamic Web Project"对话框中新建一个项目，项目名为"chapter06"，如图 6-3 所示。

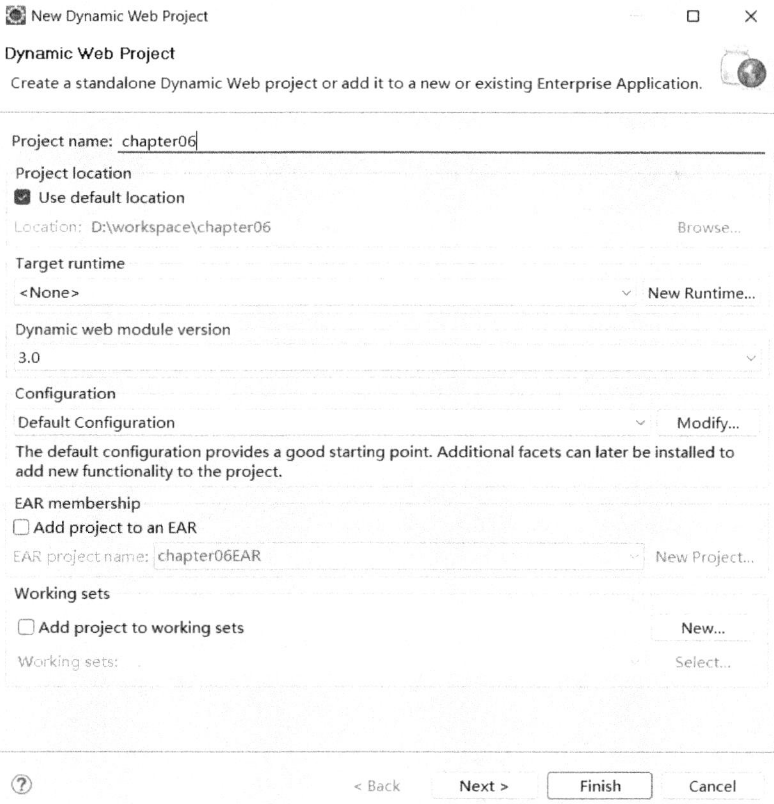

图 6-3 "New Dynamic Web Project"对话框

单击"Next"按钮，在打开的新窗口中再次单击"Next"按钮，此时打开了如图 6-4 所示的"Web Module"对话框,单击"Finish"按钮即可完成项目创建。

图 6-4 "Web Module"对话框

第二步:新建 Servlet 程序。

① 先在 chapter06 项目中展开折叠的 Java Resources 文件夹,然后右键单击 src 文件夹,依次选择"New"→"Servlet"选项,如图 6-5 所示。

图 6-5 依次选择"New"→"Servlet"选项

② 打开 "Create Servlet" 对话框，输入 Class name 的值为 "TestServlet"，如图 6-6 所示。

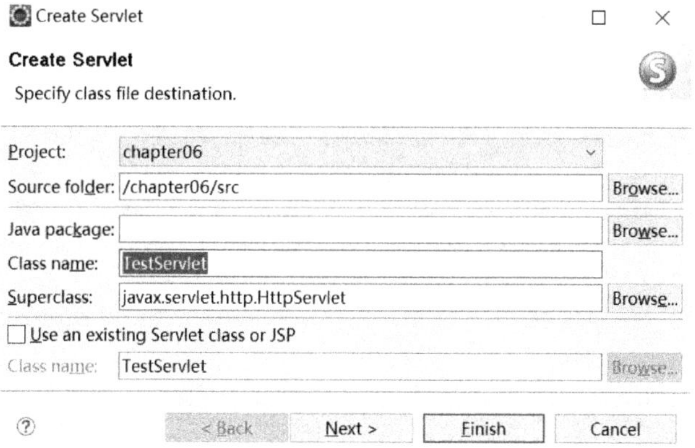

图 6-6 输入 Class name 的值

③ 在图 6-6 中单击 "Finish" 按钮，便会自动创建类名为 "TestServlet" 的 Servlet 类，它继承了 HttpServlet 类，并自动生成了 doGet()方法和 doPost()方法，分别用于处理 GET 请求和 POST 请求。

【例 6-2】配置 TestServlet 类。

```java
import java.io.IOException;
import javax.servlet.ServletException;
import javax.servlet.annotation.WebServlet;
import javax.servlet.http.HttpServlet;
import javax.servlet.http.HttpServletRequest;
import javax.servlet.http.HttpServletResponse;
/**
 * Servlet implementation class TestServlet
 */
@WebServlet("/TestServlet")
public class TestServlet extends HttpServlet {
    private static final long serialVersionUID = 1L;

    /**
     * @see HttpServlet#HttpServlet()
     */
    public TestServlet() {
        super();
        // TODO Auto-generated constructor stub
    }
    /**
     * @see HttpServlet#doGet(HttpServletRequest request,HttpServletResponse response)
     */
    protected void doGet(HttpServletRequest request,HttpServletResponse response) throws ServletException,IOException {
        // TODO Auto-generated method stub
```

```
        }
        /**
         * @see HttpServlet#doPost(HttpServletRequest request,HttpServletResponse
response)
         */
        protected void doPost(HttpServletRequest request,HttpServletResponse
response) throws ServletException,IOException {
            // TODO Auto-generated method stub
        }
    }
```

第三步：配置 Servlet 程序。

对例 6-2 中的代码 "@WebServlet（"/TestServlet"）" 进行修改，将 TestServlet 类标注为 Servlet 程序，可将该行代码改为：

```
@WebServlet(name="TestServlet",urlPatterns="/TestServlet")
```

其中，name 属性将 Servlet 程序名设置为 "TestServlet"，urlPatterns 属性用于设置与 Servlet 程序匹配的 URL（即访问该 Servlet 程序的 URL 为 "/TestServlet"）。

【例 6-3】修改 "@WebServlet（"/TestServlet"）" 后的 TestServlet 类。参考代码可扫描侧方二维码查看。

第四步：修改 doGet()方法，实现动态生成网页内容。

将 doGet()方法中的代码修改为如下内容。

```
    response.setContentType("text/html;charset=utf-8");
PrintWriter out=response.getWriter();
out.print("自主创新,中国软件从追赶到超越");
```

修改结果如下方的例 6-4 所示。

【例 6-4】在修改 doGet()方法后得到的 TestServlet 类如下所示。

```
import java.io.IOException;
import java.io.PrintWriter;
import javax.servlet.ServletException;
import javax.servlet.annotation.WebServlet;
import javax.servlet.http.HttpServlet;
import javax.servlet.http.HttpServletRequest;
import javax.servlet.http.HttpServletResponse;
/**
 * Servlet implementation class TestServlet
 */
@WebServlet(name="TestServlet",urlPatterns="/TestServlet")
public class TestServlet extends HttpServlet {
    private static final long serialVersionUID = 1L;
    /**
     * @see HttpServlet#HttpServlet()
     */
    public TestServlet() {
```

```
        super();
        // TODO Auto-generated constructor stub
    }
    /**
     * @see HttpServlet#doGet(HttpServletRequest request,HttpServletResponse response)
     */
    protected void doGet(HttpServletRequest request,HttpServletResponse response) throws ServletException,IOException {
        // TODO Auto-generated method stub
        response.setContentType("text/html;charset=utf-8");
        PrintWriter out=response.getWriter();
        out.print("自主创新,中国软件从追赶到超越");
    }
    /**
     * @see HttpServlet#doPost(HttpServletRequest request,HttpServletResponse response)
     */
    protected void doPost(HttpServletRequest request,HttpServletResponse response) throws ServletException,IOException {
        // TODO Auto-generated method stub
    }
}
```

第五步：启动 Tomcat 服务器。

① 在 Eclipse 主界面的菜单栏中，先单击"Run"菜单，然后再次单击"Run"选项，如图 6-7 所示。

图 6-7 单击"Run"菜单

② 打开如图 6-8 所示的"Run On Server"窗口，单击"Finish"按钮启动 Tomcat 服务器。Tomcat 服务器启动成功的效果如图 6-9 所示。

第六步：访问 Servlet。

在 Eclipse 主界面单击如图 6-10 所示的图标（即"Open Web Browser"按钮），进而打开如图 6-11 所示的浏览器窗口，在浏览器窗口输入 Servlet 的访问地址"http://localhost:8080/chapter06/TestServlet"，效果如图 6-12 所示。访问地址"http://localhost:8080/chapter06/TestServlet"中的"localhost:8080/chapter06"是 Web 项目的访问地址，而"/TestServlet"则

是在注解"@WebServlet（name="TestServlet",urlPatterns="/TestServlet"）"中的 urlPatterns 属性设置的访问该 Servlet 的 URL 地址信息。

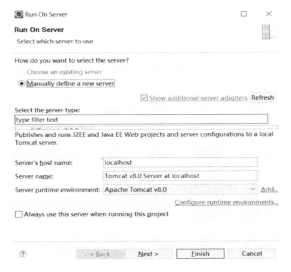

图 6-8　"Run On Server"窗口

图 6-9　Tomcat 服务器启动成功的效果

图 6-10　单击"Open Web Browser"按钮

图 6-11　浏览器窗口

图 6-12　输入访问地址的效果

任务拓展

1. ServletConfig 对象

在上文中，我们配置了 Servlet 的名称及访问的 URL 地址，用于请求访问该 Servlet，但在 Servlet 运行期间，经常用到一些配置信息，如文件使用的编码等，这些信息都可以通过相应的属性事先配置在@WebServlet 注解中，但是如何在运行期间获取并使用这些配置信息呢？我们可以通过 ServletConfig 对象实现。

当 Servlet 容器初始化 Servlet 时，该 Servlet 的配置信息会被封装到 ServletConfig 对象中，并通过调用 init(ServletConfig config)方法将 ServletConfig 对象传递给 Servlet 实例。因此，在该 Servlet 运行期间，它可以通过调用 ServletConfig 接口定义的相关方法来获取相应的配置信息。使用 ServletConfig 对象获取配置信息的方法如表 6-4 所示。

表 6-4 使用 ServletConfig 对象获取配置信息的方法

方法	功能
String getInitParameter(String name)	根据初始化参数名返回对应的初始化参数值
Enumeration getInitParameterNames()	返回一个 Enumeration 对象，该对象中包含了所有的初始化参数名
ServletContext getServletContext()	返回一个代表当前 Web 应用的 ServletContext 对象
String getServletName()	返回 Servlet 的名字

表 6-4 中的 getInitParameter()方法是获取配置信息的常用方法，请务必掌握。此外，getServletContext()方法也可以用来获取 ServletContext 对象，ServletContext 对象将在后续具体讲解。接下来，我们通过例 6-5 理解如何在 Servlet 运行期间使用 getInitParameter()方法获取配置信息。

教学视频

【例 6-5】在 Servlet 运行期间使用 getInitParameter()方法获取配置信息。

```
1  import java.io.IOException;
2  import java.io.PrintWriter;
3
4  import javax.servlet.Servlet;
5  import javax.servlet.ServletConfig;
6  import javax.servlet.ServletException;
7  import javax.servlet.annotation.WebInitParam;
8  import javax.servlet.annotation.WebServlet;
9  import javax.servlet.http.HttpServlet;
10 import javax.servlet.http.HttpServletRequest;
11 import javax.servlet.http.HttpServletResponse;
12 /**
13  * Servlet implementation class TestServlet2
14  */
15 @WebServlet(name="TestServlet2",urlPatterns="/TestServlet2",
16    initParams={@WebInitParam(name="encoding",value="UTF-8")})
18 public class TestServlet2 extends HttpServlet {
19     private static final long serialVersionUID = 1L;
20     /**
```

```
21      * @see HttpServlet#HttpServlet()
22      */
23     public TestServlet2() {
24         super();
25         // TODO Auto-generated constructor stub
26     }
27     /**
28 /    * @see HttpServlet#doGet(HttpServletRequest request,HttpServletResponse
29 response)
30      */
31     protected void doGet(HttpServletRequest request,HttpServletResponse
32 response) throws ServletException,IOException {
33         // TODO Auto-generated method stub
34         ServletConfig config=this.getServletConfig();//实例化ServletConfig对
35 象,对象名为"config"
36         String encode_Param=config.getInitParameter("encoding");//使用getInit
37 Parameter()方法获取名为encoding的参数对应的参数值
38         //输出获取的编码信息
39         PrintWriter  out=response.getWriter();
40         out.print("encoding="+encode_Param);
41     }
42     /**
43      * @see HttpServlet#doPost(HttpServletRequest request,HttpServletResponse
44 response)
45      */
46     protected void doPost(HttpServletRequest request,HttpServletResponse
47 response) throws ServletException,IOException {
48         // TODO Auto-generated method stub
49         doGet(request,response);
50     }
51 }
```

代码解析：第16行在@WebServlet注解中使用WebInitParam属性设置参数encoding的参数值为"UTF-8"，即设置文件编码为UTF-8；第34行使用getServletConfig()方法获取ServletConfig对象，其对象名为"config"；第36行通过对象名为"config"的ServletConfig对象调用getInitParameter()方法获取名为"encoding"的参数的参数值，即获取文件编码。

输出获取的文件编码，效果如图6-13所示。

```
http://localhost:8080/chapter06/TestServlet2
encoding=UTF-8
```

图6-13　输出获取的文件编码的效果

2. ServletContext对象

在一个Servlet容器（如Tomcat服务器）中，可能会同时运行多个网站（即多个Web应用），那么如何区分这些Web应用呢？我们可以使用ServletContext对象实现。Servlet容器会为每个Web应用创建一个唯一的ServletContext对象，用于表示当前的Web应用。

ServletContext 对象不仅封装了当前 Web 应用的所有信息，还能实现多个 Servlet 之间的数据共享。

（1）获取 Web 应用的初始化信息

可以在 web.xml 文件中配置整个网站（即 Web 应用）的初始化信息，如图 6-14 所示。在 web.xml 文件中配置的参数 organizer 用于表示网站的所有机构，其参数值为"Chongqing Vocational City College"。此外，还配置了参数 address 用于表示地址，其参数值为"Chongqing"。

我们可以通过 ServletContext 对象的 getInitParameterNames()方法和 getInitParameter()方法，分别获取上述参数名和参数值，下面通过例 6-6 来实现。

```xml
web.xml
1 <?xml version="1.0" encoding="UTF-8"?>
2 <!DOCTYPE web-app PUBLIC "-//Sun Microsystems, Inc.//DTD Web Application 2.3//EN" "http://java.sun.com/dtd/web-app_2_3.dtd">
3 <web-app>
4     <context-param>
5         <param-name>organizer</param-name>
6         <param-value>Chongqing Vocational City College</param-value>
7     </context-param>
8     <context-param>
9         <param-name>address</param-name>
10        <param-value>Chongqing</param-value>
11    </context-param>
12 </web-app>
```

图 6-14 配置 Web 应用的初始化信息

【例 6-6】通过 ServletContext 对象获取 Web 应用的初始化信息。

```java
1  import java.io.IOException;
2  import java.io.PrintWriter;
3  import java.util.Enumeration;
4  import javax.servlet.ServletContext;
5  import javax.servlet.ServletException;
6  import javax.servlet.annotation.WebServlet;
7  import javax.servlet.http.HttpServlet;
8  import javax.servlet.http.HttpServletRequest;
9  import javax.servlet.http.HttpServletResponse;
10 /**
11  * Servlet implementation class TestServlet3
12  */
13 @WebServlet(name="TestServlet3",urlPatterns="/TestServlet3")
14 public class TestServlet3 extends HttpServlet {
15     private static final long serialVersionUID = 1L;
16     /**
17      * @see HttpServlet#HttpServlet()
18      */
19     public TestServlet3() {
20         super();
21         // TODO Auto-generated constructor stub
22     }
23     /**
24      * @see HttpServlet#doGet(HttpServletRequest request,HttpServletResponse
25 response)
26      */
27     protected void doGet(HttpServletRequest request,HttpServletResponse
```

```
28  response) throws ServletException,IOException {
29          // TODO Auto-generated method stub
30          ServletContext context=this.getServletContext();//获取ServletContext
31  对象,对象名为"context"
32          Enumeration<String> paramNames=context.getInitParameterNames();//通过
33  对象名为"context"的ServletContext对象调用getInitParameterNames()方法获取所有的
34  初始化参数名
35                                                                  //并将其保
36  存到枚举对象paramNames中
37          String name=paramNames.nextElement();//可以使用nextElement()方法依次获
38  取paramNames中的参数,每执行一次获取一个参数
39                                                      //并将当前参数名保存到name变量中
40          String value=context.getInitParameter(name);//通过对象名为"context"的
41  ServletContext对象调用getInitParameter()方法获取name中的参数名"address"对应的的参
42  数值
43          //输出获取到的参数名及其参数值
44          PrintWriter out=response.getWriter();
45          out.print(name+":"+value);
46      }
47      /**
48       * @see HttpServlet#doPost(HttpServletRequest request,HttpServletResponse
49  response)
50       */
51      protected void doPost(HttpServletRequest request,HttpServletResponse
52  response) throws ServletException,IOException {
53          // TODO Auto-generated method stub
54      }
55  }
```

代码解析：第 30 行调用 getServletContext()方法获取 ServletContext 对象，对象名为"context"；第 32 行使用对象名为"context"的 ServletContext 对象调用 getInitParameterNames()方法，获取所有的初始化参数名，并将其保存到枚举对象 paramNames 中；第 37 行使用 nextElement()方法依次获取 paramNames 中的参数，每执行一次可获取一个参数；第 40 行通过对象名为"context"的 ServletContext 对象调用 getInitParameter()方法，获取变量 name 中的参数名"address"对应的参数值"Chongqing"，并保存到变量 value 中。执行效果如图 6-15 所示。

图 6-15　执行效果

（2）多个 Servlet 共享数据

由于 Web 应用有且只有一个 ServletContext 对象，而 ServletContext 对象的域属性又可

以被所有 Servlet 访问，因此可以通过 ServletContext 对象的域属性实现多个 Servlet 共享数据。ServletContext 对象的域属性的常用操作方法如表 6-5 所示。

表 6-5 ServletContext 对象的域属性的常用操作方法

方法	功能
Enumeration getAttributeNames()	返回一个 Enumeration 对象，该对象包含了所有存放在 ServletContext 对象中的域属性的名称
Object getAttribute(String name)	根据参数 name 指定的属性名返回一个与之匹配的域属性的值
void removeAttribute(String name)	根据参数 name 指定的域属性的名称，从 ServletContext 对象中删除与之匹配的域属性
void setAttribute(String name,Object obj)	设置 ServletContext 对象的域属性，其中参数 name 是域属性名，参数 obj 是域属性的值

【例 6-7】通过 setAttribute()方法设置 ServletContext 对象的域属性 data，其值为"This is a test data"。

教学视频

```
import java.io.IOException;
import javax.servlet.ServletContext;
import javax.servlet.ServletException;
import javax.servlet.annotation.WebServlet;
import javax.servlet.http.HttpServlet;
import javax.servlet.http.HttpServletRequest;
import javax.servlet.http.HttpServletResponse;
/**
 * Servlet implementation class TestServlet4
 */
@WebServlet(name="TestServlet4",urlPatterns="/TestServlet4")
public class TestServlet4 extends HttpServlet {
    private static final long serialVersionUID = 1L;
    /**
     * @see HttpServlet#HttpServlet()
     */
    public TestServlet4() {
        super();
        // TODO Auto-generated constructor stub
    }
    /**
     * @see HttpServlet#doGet(HttpServletRequest request,HttpServletResponse response)
     */
    protected void doGet(HttpServletRequest request,HttpServletResponse response) throws ServletException,IOException {
        // TODO Auto-generated method stub
        ServletContext context=this.getServletContext();//获取ServletContext对象,对象名为"context"
        context.setAttribute("data","This is a test data");//通过名为"context"的ServletContext对象调用setAttribute()方法创建名为"data"的域属性
```

```
            //其值为"This is a test data"
        }
        /**
         * @see HttpServlet#doPost(HttpServletRequest request,HttpServletResponse response)
         */
        protected void doPost(HttpServletRequest request,HttpServletResponse response) throws ServletException,IOException {
            // TODO Auto-generated method stub
        }
    }
```

【例 6-8】使用 getAttribute()方法获取 ServletContext 对象的 data 域属性的值，执行结果如图 6-16 所示。

```
import java.io.IOException;
import java.io.PrintWriter;
import javax.servlet.ServletContext;
import javax.servlet.ServletException;
import javax.servlet.annotation.WebServlet;
import javax.servlet.http.HttpServlet;
import javax.servlet.http.HttpServletRequest;
import javax.servlet.http.HttpServletResponse;

/**
 * Servlet implementation class TestServlet5
 */
@WebServlet(name="TestServlet5",urlPatterns="/TestServlet5")
public class TestServlet5 extends HttpServlet {
    private static final long serialVersionUID = 1L;
    /**
     * @see HttpServlet#HttpServlet()
     */
    public TestServlet5() {
        super();
        // TODO Auto-generated constructor stub
    }
    /**
     * @see HttpServlet#doGet(HttpServletRequest request,HttpServletResponse response)
     */
    protected void doGet(HttpServletRequest request,HttpServletResponse response) throws ServletException,IOException {
        // TODO Auto-generated method stub
        ServletContext context=this.getServletContext();//获取 ServletContext 对象,对象名为"context"
        String data=(String)context.getAttribute("data");//使用 getAttibute()方法获取 ServletContext 对象名为"data"的域属性的值
```

```
        //输出获取的属性值                    //并将其保存到变量data中
        PrintWriter  out=response.getWriter();
        out.print(data);
    }
    /**
     * @see HttpServlet#doPost(HttpServletRequest  request,HttpServletResponse
response)
     */
    protected void doPost(HttpServletRequest request,HttpServletResponse response)
throws ServletException,IOException {
        // TODO Auto-generated method stub
    }}
```

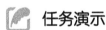

图 6-16 设置域属性的执行结果

任务 2 应用 HttpServletRequest 对象和 HttpServletResponse 对象实现用户验证登录

任务演示

2022 年 6 月 1 日刊发的《〈网络安全法〉五周年：网络安全为人民 网络安全靠人民》中提到："作为我国网络安全领域的基础性法律，《网络安全法》是我国网络空间法治建设的重要里程碑"。对于动态网站来说，用户登录时的账号验证、密码验证都是验证用户身份，保障网络安全的基本措施。本次任务我们将使用 HttpServletRequest 对象和 HttpServletResponse 对象实现用户的验证登录，登录界面、验证登录成功界面分别如图 6-17、图 6-18 所示。

图 6-17 登录界面

图 6-18 验证登录成功界面

知识准备

1. HttpServletRequest 对象

HttpServletRequest 接口继承自 ServletRequest 接口，用于封装客户端的请求信息，例如，客户端将用户名、密码传送过来，请求访问某个 Servlet 进行验证和登录，其请求信息

（用户名、密码等）就会被封装成 HttpServletRequest 对象。当不同的请求信息被封装到 HttpServletRequest 对象后，该如何获取呢？我们可以通过 HttpServletRequest 对象的不同种类的方法进行获取。

（1）获取请求行信息的相关方法

当客户端发送请求信息访问或调用某个 Servlet 程序时，该请求信息的请求行中包含了请求方法、请求资源名、请求路径等信息。为获取这些请求行信息，我们可以使用如表 6-6 所示的获取请求行信息的相关方法。

表 6-6　获取请求行信息的相关方法

方法	功能
String getMethod()	该方法用于获取请求信息中的请求方法（如 GET 请求、POST 请求等）
String getRequestURI()	该方法用于获取请求行中资源名的部分，即位于 URL 的主机和端口之后、参数之前的数据
String getQueryString()	该方法用于获取请求行中的参数部分，也就是请求路径后面问号（?）以后的所有内容
String getProtocol()	该方法用于获取请求行中的协议名和版本，例如 HTTP/1.0 或 HTTP/1.1
String getContextPath()	该方法用于获取请求 URL 中属于 Web 应用程序的路径，这个路径以"/"开头，表示相对于整个 Web 站点的根目录，路径结尾不含"/"。如果请求 URL 属于 Web 站点的根目录，那么返回结果为空字符串（""）
String getServletPath()	该方法用于获取 Servlet 的名称或 Servlet 映射的路径
String getRemoteAddr()	该方法用于获取请求客户端的 IP 地址，其格式类似于"192.168.0.3"
String getRemoteHost()	该方法用于获取请求客户端的完整主机名，其格式类似于"pc1.itcast.cn"。需要注意的是，如果无法解析出客户端的完整主机名，该方法将会返回客户端的 IP 地址
int getRemotePort()	该方法用于获取请求客户端网络连接的端口号
String getLocalAddr()	该方法用于获取 Web 服务器上接收当前请求网络连接的 IP 地址
String getLocalName()	该方法用于获取 Web 服务器上接收当前网络连接 IP 所对应的主机名
int getLocalPort()	该方法用于获取 Web 服务器上接收当前网络连接的端口号
String getServerName()	该方法用于获取当前请求指向的主机名，即请求信息中 host 头字段对应的主机名部分
int getServerPort()	该方法用于获取当前请求连接的服务器端口号，即请求信息中 host 头字段对应的端口号部分
String getScheme()	该方法用于获取请求的协议名，例如 HTTP、HTTPS 或 FTP
StringBuffer getRequestURL()	该方法用于获取客户端发出请求时的完整 URL，包括协议、服务器名、端口号、请求路径等信息，但不包括后面的查询参数部分。注意，getRequestURL()方法返回的结果是 StringBuffer 类型的，而不是 String 类型的，这样更便于对结果进行修改

为了便于理解各种获取请求行信息的相关方法，例 6-9 的 TestServlet6 演示了如何使用上述方法。

【例 6-9】获取请求行信息的相关方法的应用示例，参考代码如下，运行结果如图 6-19 所示。

```
import java.io.IOException;
import java.io.PrintWriter;
import javax.servlet.ServletException;
import javax.servlet.annotation.WebServlet;
import javax.servlet.http.HttpServlet;
```

```java
import javax.servlet.http.HttpServletRequest;
import javax.servlet.http.HttpServletResponse;
/**
 * Servlet implementation class TestServlet6
 */
@WebServlet(name="TestServlet6",urlPatterns="/TestServlet6")
public class TestServlet6 extends HttpServlet {
    private static final long serialVersionUID = 1L;
    /**
     * @see HttpServlet#HttpServlet()
     */
    public TestServlet6() {
        super();
        // TODO Auto-generated constructor stub
    }
    /**
     * @see HttpServlet#doGet(HttpServletRequest request,HttpServletResponse response)
     */
    protected void doGet(HttpServletRequest request,HttpServletResponse response) throws ServletException,IOException {
        // TODO Auto-generated method stub
        PrintWriter out = response.getWriter();
        // 获取请求行的相关信息
        out.println("getMethod:" + request.getMethod() + "<br/>");
        out.println("getRequestURI:" + request.getRequestURI() + "<br/>");
        out.println("getQueryString:"+request.getQueryString() + "<br/>");
        out.println("getProtocol:" + request.getProtocol() + "<br/>");
        out.println("getContextPath:"+request.getContextPath() + "<br/>");
        out.println("getPathInfo:" + request.getPathInfo() + "<br/>");
        out.println("getPathTranslated:"+ request.getPathTranslated() + "<br/>");
        out.println("getServletPath:"+request.getServletPath() + "<br/>");
        out.println("getRemoteAddr:" + request.getRemoteAddr() + "<br/>");
        out.println("getRemoteHost:" + request.getRemoteHost() + "<br/>");
        out.println("getRemotePort:" + request.getRemotePort() + "<br/>");
        out.println("getLocalAddr:" + request.getLocalAddr() + "<br/>");
        out.println("getLocalName:" + request.getLocalName() + "<br/>");
        out.println("getLocalPort:" + request.getLocalPort() + "<br/>");
        out.println("getServerName:" + request.getServerName() + "<br/>");
        out.println("getServerPort:" + request.getServerPort() + "<br/>");
        out.println("getScheme:" + request.getScheme() + "<br/>");
        out.println("getRequestURL:" + request.getRequestURL() + "<br/>");
    }
    /**
     * @see HttpServlet#doPost(HttpServletRequest request,HttpServletResponse response)
     */
```

```
        protected void doPost(HttpServletRequest request,HttpServletResponse response)
throws ServletException,IOException {
            // TODO Auto-generated method stub
        }
    }
```

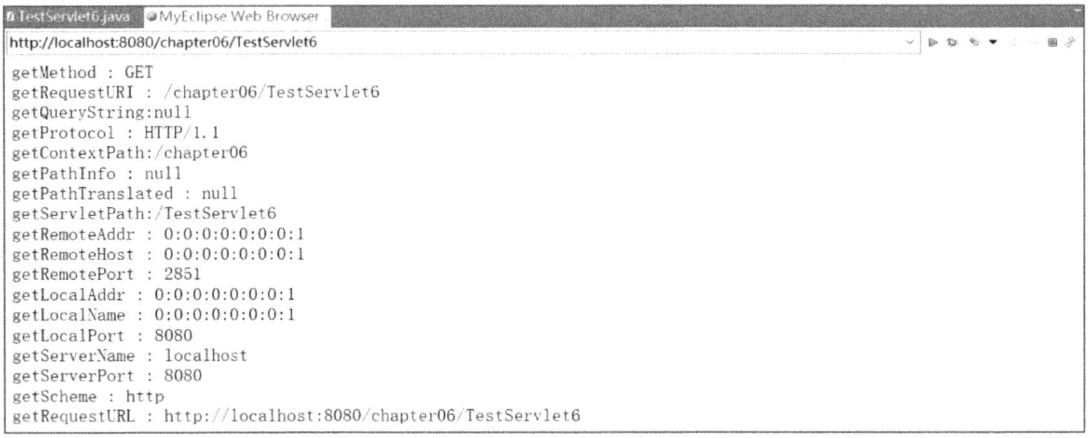
TestServlet6.java
MyEclipse Web Browser
http://localhost:8080/chapter06/TestServlet6
getMethod : GET
getRequestURI : /chapter06/TestServlet6
getQueryString:null
getProtocol : HTTP/1.1
getContextPath:/chapter06
getPathInfo : null
getPathTranslated : null
getServletPath:/TestServlet6
getRemoteAddr : 0:0:0:0:0:0:0:1
getRemoteHost : 0:0:0:0:0:0:0:1
getRemotePort : 2851
getLocalAddr : 0:0:0:0:0:0:0:1
getLocalName : 0:0:0:0:0:0:0:1
getLocalPort : 8080
getServerName : localhost
getServerPort : 8080
getScheme : http
getRequestURL : http://localhost:8080/chapter06/TestServlet6

图 6-19 从 HttpServletRequest 对象中获取请求行信息的运行结果

（2）获取请求头的相关方法

当客户端请求访问某个 Servlet 时，请求头可以向服务器传送附加信息，如客户端可以接收的数据类型、语言等。那么如何从 HttpServletRequest 对象中获取这些请求头信息呢？可以通过如表 6-7 所示的获取请求头信息的相关方法实现。

表 6-7 获取请求头信息的相关方法

方法	功能
String getHeader(String name)	该方法用于获取一个指定头字段的值，如果请求信息中没有包含指定的头字段，则 getHeader()方法返回 null；如果请求信息中包含多个指定名称的头字段，则 getHeader()方法返回其中的第一个头字段的值
Enumeration getHeaders(String name)	该方法返回一个 Enumeration 对象，该集合对象由请求信息中出现的某个指定名称的所有头字段值组成。在多数情况下，一个头字段名在请求信息中只出现一次，但有时候可能会出现多次
Enumeration getHeaderNames()	该方法用于获取一个包含所有请求头字段的 Enumeration 对象
int getIntHeader(String name)	该方法用于获取指定名称的头字段，并且将其值转为整型值。需要注意的是，如果指定名称的头字段不存在，返回值为-1；如果获取到的头字段的值不能转为整型值，将发生 NumberFormatException 异常
long getDateHeader(String name)	该方法用于获取指定头字段的值，并将其按 GMT 时间格式转换成一个代表日期/时间的长整数，这个长整数是自 1970 年 1 月 1 日 0 点 0 分 0 秒起算的、以毫秒为单位的时间值
String getContentType()	该方法用于获取 Content-Type 头字段的值，结果为字符串类型值
int getContentLength()	该方法用于获取 Content-Length 头字段的值，结果为整型值
String getCharacterEncoding()	该方法用于返回请求信息的实体部分的字符集编码，通常从 Content-Type 头字段中进行提取，结果为字符串类型值

为了更好地掌握这些方法，我们在 TestServlet7 中以 getHeader()方法为例，演示如何使用 getHeader()方法获取客户端浏览器支持语言的请求头信息，如例 6-10 所示。

【例 6-10】 使用 getHeader()方法获取客户端浏览器支持语言的请求头信息，运行结果如图 6-20 所示。

```java
import java.io.IOException;
import java.io.PrintWriter;
import javax.servlet.ServletException;
import javax.servlet.annotation.WebServlet;
import javax.servlet.http.HttpServlet;
import javax.servlet.http.HttpServletRequest;
import javax.servlet.http.HttpServletResponse;
/**
 * Servlet implementation class TestServlet7
 */
@WebServlet(name="TestServlet7",urlPatterns="/TestServlet7")
public class TestServlet7 extends HttpServlet {
    private static final long serialVersionUID = 1L;
    /**
     * @see HttpServlet#HttpServlet()
     */
    public TestServlet7() {
        super();
        // TODO Auto-generated constructor stub
    }
    /**
     * @see HttpServlet#doGet(HttpServletRequest request,HttpServletResponse response)
     */
    protected void doGet(HttpServletRequest request,HttpServletResponse response) throws ServletException,IOException {
        // TODO Auto-generated method stub
        String accept_Language=request.getHeader("accept-language");//通过对象名为 request 的 HttpServletRequest 对象调用 getHeader()方法
                                                                    //获取字段 accept_language 的值,即客户端浏览器支持的语言
        PrintWriter out=response.getWriter();
        out.print(accept_Language);
    }
    /**
     * @see HttpServlet#doPost(HttpServletRequest request,HttpServletResponse response)
     */
    protected void doPost(HttpServletRequest request,HttpServletResponse response) throws ServletException,IOException {
        // TODO Auto-generated method stub
    }
}
```

图 6-20 使用 getHeader()方法获取支持语言的请求头信息的运行结果

（3）获取请求参数

在动态网站的实际运行过程中，经常会收到用户提交的各种表单数据，如登录时用户提交的用户名、密码等。HttpServletRequest 对象提供了一系列用于获取表单中的请求参数的方法，如表 6-8 所示。

表 6-8 获取表单中的请求参数的方法

方法	功能
String getParameter(String name)	该方法用于获取某个指定名称的参数的值，如果请求信息没有包含指定名称的参数，则 getParameter()方法返回 null；如果存在指定名称的参数，但没有设置该参数的值，则返回一个空字符串；如果请求信息包含多个指定名称的参数，则 getParameter()方法返回第一个出现的参数的值
String[] getParameterValues(String name)	该方法用于返回一个字符串类型的数组，请求信息可以有多个相同名称的参数（通常由一个包含多个同名的字段元素的 form 表单生成），如果要获得请求信息中的同一个参数名对应的所有参数值，那么就应该使用 getParameterValues()方法
Enumeration getParameterNames()	该方法用于返回一个包含请求信息中所有参数名的 Enumeration 对象，在此基础上，可以对请求信息中的所有参数进行遍历处理
Map getParameterMap()	该方法用于将请求信息中的所有参数名和参数值装入一个 Map 对象中，并返回该对象

在实际应用中，getParameter()和 getParameterValues()是最基本的获取请求参数的方法，其中，getParameter()方法用于获取某个指定名称的参数值（如提交的用户名、密码等），而 getParameterValues()方法用于获取多个同名的参数，如提交的用户爱好信息，因为用户可能有多个爱好，在实际的页面上会用多个同名的复选框供用户选择，所以获取多个同名参数就必须使用 getParameterValues()方法。如图 6-21 所示的用户注册页面，其对应的网页源代码如例 6-11 所示。

图 6-21 用户注册页面

教学视频

【例 6-11】编写用户注册页面，源代码如下所示。

```
1 <%@ page language="java" import="java.util.*" pageEncoding="utf-8"%>
2 <!DOCTYPE HTML PUBLIC "-//W3C//DTD HTML 4.01 Transitional//EN">
3 <html>
4   <head>
5     <title>用户注册页面</title>
6   </head>
```

```
7  <body>
8    <form action="/chapter06/TestServlet8" method="POST">
9        用户名:<input type="text" name="username"><br/>
10       密    码:<input type="password" name="password"> <br/>
11       爱好:
12       <input type="checkbox" name="hobby" value="basketball">篮球
13       <input type="checkbox" name="hobby" value="football">足球
14       <input type="checkbox" name="hobby" value="volleyball">排球
15       <input type="checkbox" name="hobby" value="badminton">羽毛球<br/>
16       <input type="submit" value="提交">
17   </form>
18 </body>
19 </html>
```

在例 6-11 中,我们可以看到"用户名""密码"的参数名(即文本框、密码框的 name 属性)是唯一的,而"爱好"的参数名(即复选框的 name 属性)有 4 个(即多个同名参数),因此"用户名""密码"请求参数的获取可以使用 getParameter()方法,而"爱好"请求参数的获取需要使用 getParameterValues()方法。注意,在例 6-11 的第 8 行的 action 属性中,明确指明了该表单的相关注册信息将会提交给 TestServlet8 进行处理,TestServlet8 的代码如例 6-12 所示。

【例 6-12】编写处理注册信息的 Servlet 程序,代码如下。

```
1  import java.io.IOException;
2  import java.io.PrintWriter;
3  import javax.servlet.ServletException;
4  import javax.servlet.annotation.WebServlet;
5  import javax.servlet.http.HttpServlet;
6  import javax.servlet.http.HttpServletRequest;
7  import javax.servlet.http.HttpServletResponse;
8  /**
9   * Servlet implementation class TestServlet8
10  */
11 @WebServlet(name="TestServlet8",urlPatterns="/TestServlet8")
12 public class TestServlet8 extends HttpServlet {
13     private static final long serialVersionUID = 1L;
14     public TestServlet8() {
15         super();
16         // TODO Auto-generated constructor stub
17     }
18     /**
19      * @see HttpServlet#doGet(HttpServletRequest request,HttpServletResponse
20 response)
21      */
22     protected void doGet(HttpServletRequest request,HttpServletResponse
23 response) throws ServletException,IOException {
24         // TODO Auto-generated method stub
25     }
```

```
26     /**
27      * @see HttpServlet#doPost(HttpServletRequest request,HttpServletResponse
28 response)
29      */
30     protected void doPost(HttpServletRequest request,HttpServletResponse
31 response) throws ServletException,IOException {
32         // TODO Auto-generated method stub
33         String name = request.getParameter("username");//使用getParameter()方法获
34 取参数名为"username"的请求参数值
35         String password = request.getParameter("password");//使用getParameter()方
36 法获取参数名为"password"的请求参数值
37         String[] hobbys = request.getParameterValues("hobby");//使用getParameterValues()
38 方法获取所有参数名为"hobby"的请求参数值
39         //输出获取的请求参数值
40         response.setContentType("text/html;charset=utf-8");
41         PrintWriter  out=response.getWriter();
42         out.print("用户名:"+name+"</br>");
43         out.print("密  码:"+password+"</br>");
44         String hobby="";
45         for (int i = 0;i < hobbys.length;i++) {
46             hobby=hobby+ "    "+hobbys[i];
47         }
48         out.print("爱  好:"+hobby+"</br>");
49     }
50 }
```

代码解析：第33行、第35行分别使用getParameter()方法获取参数名为"username"和"password"的请求参数值；第37行使用getParameterValues()方法获取所有参数名为"hobby"的请求参数值，生成的动态网页内容如图6-22所示。

图6-22　生成的动态网页内容

注意：上例中的用户名使用的是英文格式的，如果输入汉字格式的用户名则会出现中文乱码问题，注册汉字格式的用户名界面和用户名显示乱码界面分别如图6-23和图6-24所示。

图6-23　注册汉字格式的用户名界面

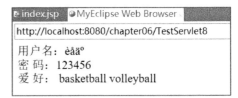

图6-24　用户名显示乱码界面

在如图 6-24 所示的动态生成的网页内容中,用户名部分显示的是乱码。针对这种中文乱码问题,HttpServletRequest 对象提供了 setCharacterEncoding()方法设置 request 对象的解码方式,显示中文使用 UTF-8。

【例 6-13】使用 setCharacterEncoding()方法设置 request 对象的解码方式为 UTF-8,运行结果如图 6-25 所示。参考代码可扫描侧方二维码查看。

例 6-13 代码

图 6-25 设置 request 对象的解码方式为 UTF-8

2. HttpServletResponse 对象

Servlet 接收客户端的请求信息并进行处理,生成并返回给客户端的响应结果由 HttpServletResponse 对象进行封装。由于返回客户端的响应结果(即 HTTP 响应消息)包括响应状态行、响应头字段、响应消息体三部分,因此 HttpServletResponse 对象也定义了向客户端发送响应状态码、响应头字段、响应消息体的方法。

(1)发送响应状态码的方法

当 Servlet 向客户端返回 HTTP 响应消息时,需要在 HTTP 响应消息中设置响应状态代码,响应状态代码代表客户端请求服务器的结果,如状态代码 200 表示处理成功,状态代码 404 表示没有找到客户端请求的资源。HttpServletResponse 对象定义了 3 个发送响应状态代码的方法,如表 6-9 所示。

表 6-9 发送响应状态代码的方法

方法	功能
setStatus(int status)	用于设置 HTTP 响应消息的状态代码,并生成响应状态行。由于响应状态行中的状态描述信息直接与状态代码相关,而 HTTP 版本由服务器确定,所以只要通过 setStatus(int status)方法设置状态代码,即可实现状态行的发送。例如,正常情况下,Web 服务器会默认产生一个状态代码为 200 的状态行
sendError(int sc)	用于发送表示错误信息的状态代码,例如,状态代码 404 表示找不到客户端请求的资源
sendError (int code,String message)	该方法除了设置状态代码,还会向客户端发出一条错误信息。服务器默认会创建一个 HTML 格式的错误服务页面作为响应结果,其中包含参数 message 指定的文本信息,这个 HTML 页面的内容类型为"text/html",保留 Cookies 和其他未修改的响应头信息。如果一个对应传入的错误码的错误页面已经在 web.xml 中声明,那么这个声明的错误页面会优先将建议的 message 参数服务于客户端

(2)发送响应头字段的方法

响应头字段用于告知客户端返回的 HTTP 响应消息的实体内容大小、返回的内容是图像还是其他类型的数据、输出内容的字符编码等。由于 HTTP 协议的响应头字段有很多种,因此 HttpServletResponse 对象也定义了一系列设置响应头字段的方法,如表 6-10 所示。

(3)发送响应消息体的方法

返回客户端的 HTTP 响应消息的响应消息体包含了大量数据,鉴于此,HttpServletResponse 对象定义了两个与输出流相关的方法用于传输大量数据,如表 6-11 所示。

表 6-10 设置响应头字段的方法

方法	功能
void addHeader(String name,String value) void setHeader(String name,String value)	这两个方法都是用来设置 HTTP 协议的响应头字段的，其中，参数 name 用于指定响应头字段的名称，参数 value 用于指定响应头字段的值。不同的是，addHeader()方法可以增加同名的响应头字段，而 setHeader()方法则会覆盖同名的响应头字段
void addIntHeader(String name, int value) void setIntHeader(String name, int value)	这两个方法专门用于设置包含整数值的响应头。避免了在调用 addHeader()方法与 setHeader()方法时，需要将 int 类型的值转换为 String 类型的值的麻烦
void setContentLength(int len)	该方法用于设置 HTTP 响应消息的实体内容的大小，单位为字节。对于 HTTP 协议来说，这个方法就是设置 Content-Length 响应头字段的值
void setContentType(String type)	该方法用于设置 Servlet 输出内容的 MIME 类型，对于 HTTP 协议来说，就是设置 Content-Type 响应头字段的值。例如，如果发送到客户端的内容是 jpeg 格式的图像数据，就需要将响应头字段的类型设置为"image/jpeg"。需要注意的是，如果响应的内容为文本，setContentType()方法还可以设置字符编码，如：text/html;charset=UTF-8
void setLocale(Locale loc)	该方法用于设置 HTTP 响应消息的本地化信息。对 HTTP 协议来说，就是设置 Content-Language 响应头字段和 Content-Type 响应头字段中的字符集编码部分。需要注意的是，如果 HTTP 响应消息没有设置 Content-Type 响应头字段，则 setLocale()方法设置的字符集编码不会出现在响应头字段中。如果调用 setCharacterEncoding()方法或 setContentType()方法指定了响应内容的字符集编码，setLocale()方法将不再具有指定字符集编码的功能
void setCharacterEncoding(String charset)	该方法用于设置输出内容使用的字符编码，对 HTTP 协议来说，就是设置 Content-Type 响应头字段中的字符集编码部分。如果没有设置 Content-Type 响应头字段，setCharacterEncoding()方法设置的字符集编码不会出现在响应头字段中。setCharacterEncoding()方法比 setContentType()方法和 setLocale()方法的优先权高，setCharacterEncoding()方法的设置结果将覆盖 setContentType()方法和 setLocale()方法

表 6-11 发送响应消息体的方法

方法	功能
getOutputStream()	该方法获取的字节输出流对象为 ServletOutputStream 类型。由于 ServletOutputStream 是 OutputStream 的子类，它可以直接输出字节数组中的二进制数据。所以，要想输出二进制格式的响应消息体，就需要调用 getOutputStream()方法
getWriter()	该方法所获取的字符输出流对象为 PrintWriter 类型。由于 PrintWriter 类型的对象可以直接输出字符文本内容，因此要想输出内容为字符文本的网页文档，需要调用 getWriter()方法

假设 Servlet 程序处理 HTTP 请求后，返回给客户端的响应消息体是"Hello, Chongqing City Vocational College！"，我们分别使用上述两个方法实现响应消息体的传输，代码分别如例 6-14、例 6-15 所示，执行结果分别如图 6-26、图 6-27 所示。

【例 6-14】使用 getOutputStream()实现响应消息体的传输。

```
import java.io.IOException;
import java.io.OutputStream;
import javax.servlet.ServletException;
import javax.servlet.annotation.WebServlet;
import javax.servlet.http.HttpServlet;
import javax.servlet.http.HttpServletRequest;
import javax.servlet.http.HttpServletResponse;
/**
```

```java
 * Servlet implementation class TestServlet9
 */
@WebServlet(name="TestServlet9",urlPatterns="/TestServlet9")
public class TestServlet9 extends HttpServlet {
    private static final long serialVersionUID = 1L;
    /**
     * @see HttpServlet#HttpServlet()
     */
    public TestServlet9() {
        super();
        // TODO Auto-generated constructor stub
    }
    /**
     * @see HttpServlet#doGet(HttpServletRequest request,HttpServletResponse response)
     */
    protected void doGet(HttpServletRequest request,HttpServletResponse response) throws ServletException,IOException {
        // TODO Auto-generated method stub
        String data="Hello,Chongqing City Vocational College!";
        OutputStream out=response.getOutputStream();//获取字节流输出对象,对象名为 out
        out.write(data.getBytes());//首先用 getBytes()方法将 data 字符串转为字节数组,
        //对象名为 out 的字节流输出对象通过 write()方法输出字节数组
    }
    /**
     * @see HttpServlet#doPost(HttpServletRequest request,HttpServletResponse response)
     */
    protected void doPost(HttpServletRequest request,HttpServletResponse response) throws ServletException,IOException {
        // TODO Auto-generated method stub
    }
}
```

【例 6-15】使用 getWriter()实现响应消息体的传输。

```java
import java.io.IOException;
import java.io.PrintWriter;
import javax.servlet.ServletException;
import javax.servlet.annotation.WebServlet;
import javax.servlet.http.HttpServlet;
import javax.servlet.http.HttpServletRequest;
import javax.servlet.http.HttpServletResponse;
/**
 * Servlet implementation class TestServlet10
 */
@WebServlet("/TestServlet10")
public class TestServlet10 extends HttpServlet {
```

```java
        private static final long serialVersionUID = 1L;
    /**
     * @see HttpServlet#HttpServlet()
     */
    public TestServlet10() {
        super();
        // TODO Auto-generated constructor stub
    }
    /**
     * @see HttpServlet#doGet(HttpServletRequest request,HttpServletResponse response)
     */
    protected void doGet(HttpServletRequest request,HttpServletResponse response) throws ServletException,IOException {
        // TODO Auto-generated method stub
        String data="Hello,Chongqing City Vocational College!";
        PrintWriter out=response.getWriter();//获取字符输出流对象,对象名为 out
        out.write(data);//对象名为 out 的字符输出流对象调用 write()方法输出 data 字符串
    }
    /**
     * @see HttpServlet#doPost(HttpServletRequest request,HttpServletResponse response)
     */
    protected void doPost(HttpServletRequest request,HttpServletResponse response) throws ServletException,IOException {
        // TODO Auto-generated method stub
    }
}
```

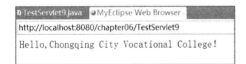

图 6-26　使用 getOutputStream()的运行结果　　　图 6-27　使用 getWriter()的运行结果

3. 请求重定向

假设一名用户没有注册账户,首先要在客户端打开注册页面填写注册信息,填写完毕后,再单击"提交"按钮将注册信息提交给 Servlet 程序处理。当 Servlet 程序处理完注册信息后,我们希望客户端的浏览器直接跳转到登录界面,以便填写账户信息进行验证、登录。但是该 Servlet 程序只有处理注册信息的功能,没有处理登录信息的功能,因此,该 Servlet 程序可以在处理完注册信息后,通过返回响应消息告知客户端一个新的资源路径(如登录网页的 URL 等),由客户端的浏览器根据新的资源路径,重新向服务器发送 HTTP 请求访问登录页面。上述过程就是一个请求重定向的过程,即 Web 服务器接收到客户端的请求后,可能由于某些条件限制,不能访问当前请求 URL 所指向的 Web 资源,而是指定了一个新的资源路径,让客户端重新发送请求。为了实现请求重定向,HttpServletResponse 接口定义了一个 sendRedirect()方法,该方法用于生成 302 响应代码和 Location 响应头字段,从而通知客户端重新访问 Location 响应头字段中指定的 URL。

【例 6-16】在如图 6-28 所示的用户注册界面中填写注册信息，单击"提交"按钮后，使用 sendRedirect()方法实现重定向，重定向的实现代码如下所示。

图 6-28　用户注册界面

教学视频

```java
import java.io.IOException;
import javax.servlet.ServletException;
import javax.servlet.annotation.WebServlet;
import javax.servlet.http.HttpServlet;
import javax.servlet.http.HttpServletRequest;
import javax.servlet.http.HttpServletResponse;
/**
 * Servlet implementation class TestServlet11
 */
@WebServlet("/TestServlet11")
public class TestServlet11 extends HttpServlet {
    private static final long serialVersionUID = 1L;

    /**
     * @see HttpServlet#HttpServlet()
     */
    public TestServlet11() {
        super();
        // TODO Auto-generated constructor stub
    }
    /**
     * @see HttpServlet#doGet(HttpServletRequest request,HttpServletResponse response)
     */
    protected void doGet(HttpServletRequest request,HttpServletResponse response) throws ServletException,IOException {
        // TODO Auto-generated method stub
    }
    /**
     * @see HttpServlet#doPost(HttpServletRequest request,HttpServletResponse response)
     */
    protected void doPost(HttpServletRequest request,HttpServletResponse response) throws ServletException,IOException {
        // TODO Auto-generated method stub
        request.setCharacterEncoding("utf-8");
```

```
            String name = request.getParameter("username");//使用 getParameter()方法
获取参数名为 username 的请求参数值
            String password = request.getParameter("password");//使用 getParameter ()
方法获取参数名为 password 的请求参数值
            String[] hobbys = request.getParameterValues("hobby");// 使 用
getParameterValues()方法获取所有参数名为 hobby 的请求参数值
            //重定向到登录界面 Login.jsp
            response.sendRedirect("/chapter06/Login.jsp");
        }
    }
```

【例 6-17】编写注册界面源代码。

```
<%@ page language="java" import="java.util.*" pageEncoding="utf-8"%>
<!DOCTYPE HTML PUBLIC "-//W3C//DTD HTML 4.01 Transitional//EN">
<html>
  <head>
    <title>用户注册页面</title>
  </head>
  <body>
    <form action="/chapter06/TestServlet11" method="POST">
        用户名:<input type="text" name="username"><br/>
        密      码 :<input type="password" name="password"><br/>
        爱好:
        <input type="checkbox" name="hobby" value="basketball">篮球
        <input type="checkbox" name="hobby" value="football">足球
        <input type="checkbox" name="hobby" value="volleyball">排球
        <input type="checkbox" name="hobby" value="badminton">羽毛球<br/>
        <input type="submit" value="提交">
    </form>
  </body>
</html>
```

【例 6-18】编写登录界面源代码。

```
<%@ page language="java" import="java.util.*" pageEncoding="utf-8"%>
<!DOCTYPE HTML PUBLIC "-//W3C//DTD HTML 4.01 Transitional//EN">
<html>
  <head>
    <title>用户登录页面</title>
  </head>
  <body>
    <form action="/chapter06/TestServlet12" method="POST">
        用户名:<input type="text" name="username"><br/>
        密      码 :<input type="password" name="password"><br/>
        <input type="submit" value="提交">
              
```

```
            <input type="Reset" value="取消">
    </form>
  </body>
</html>
```

重定向到登录界面的运行结果如图 6-29 所示。

图 6-29　重定向到登录界面的运行结果

 任务实施

任务要求：

① 用户通过 Login.jsp 提交用户名、密码。
② 检测用户名是否为"fan"，密码是否为"123456"。
③ 若用户名和密码正确，则重定位到登录成功界面，否则重定位到登录界面。

第一步：编写 Login.jsp 的源代码，用于输入并提交用户名、密码，如例 6-19 所示。

【例 6-19】编写 Login.jsp 的源代码，可扫描侧方二维码查看。

第二步：编写 Welcome.jsp 的源代码，用于实现登录成功的重定向，如例 6-20 所示。

【例 6-20】编写 Welcome.jsp 的源代码，可扫描侧方二维码查看。

第三步：编写 TestServlet12 的源代码，用于用户名、密码的验证及重定向，如例 6-21 所示。

【例 6-21】编写 TestServlet12 的源代码，可扫描侧方二维码查看。

用户登录界面及登录成功界面分别如图 6-30、图 6-31 所示。

教学视频

例 6-19 代码

例 6-20 代码

例 6-21 代码

图 6-30　用户登录界面

图 6-31　登录成功界面

 任务拓展

1. 解决中文乱码问题

我们学习了如何解决中文乱码问题（即由客户端提交到服务器的数据的编码、解码的码表不一致而导致的问题），同理，HTTP 响应消息从服务器输出到客户端，也会遇到类似的问题。下面通过具体实例进行说明。

【例 6-22】将中文字符串"我热爱编程!"输出到客户端。

```java
import java.io.IOException;
import java.io.PrintWriter;
import javax.servlet.ServletException;
import javax.servlet.annotation.WebServlet;
import javax.servlet.http.HttpServlet;
import javax.servlet.http.HttpServletRequest;
import javax.servlet.http.HttpServletResponse;
/**
 * Servlet implementation class TestServlet13
 */
@WebServlet("/TestServlet13")
public class TestServlet13 extends HttpServlet {
    private static final long serialVersionUID = 1L;
    /**
     * @see HttpServlet#HttpServlet()
     */
    public TestServlet13() {
        super();
        // TODO Auto-generated constructor stub
    }
    /**
     * @see HttpServlet#doGet(HttpServletRequest request,HttpServletResponse response)
     */
    protected void doGet(HttpServletRequest request,HttpServletResponse response) throws ServletException,IOException {
        // TODO Auto-generated method stub
        String data = "我热爱编程!";
        // 获取字符输出流对象
        PrintWriter print = response.getWriter();
        print.write(data);// 输出信息
    }
    /**
     * @see HttpServlet#doPost(HttpServletRequest request,HttpServletResponse response)
     */
    protected void doPost(HttpServletRequest request,HttpServletResponse response) throws ServletException,IOException {
        // TODO Auto-generated method stub
    }
}
```

运行结果如图 6-32 所示,从图中可以看出,显示结果是乱码。

图 6-32 显示结果是乱码

【例 6-23】使用 setContentType()方法解决例 6-22 中出现的中文乱码问题。

```java
import java.io.IOException;
import java.io.PrintWriter;
import javax.servlet.ServletException;
import javax.servlet.annotation.WebServlet;
import javax.servlet.http.HttpServlet;
import javax.servlet.http.HttpServletRequest;
import javax.servlet.http.HttpServletResponse;
/**
 * Servlet implementation class TestServlet13
 */
@WebServlet("/TestServlet13")
public class TestServlet13 extends HttpServlet {
    private static final long serialVersionUID = 1L;
    /**
     * @see HttpServlet#HttpServlet()
     */
    public TestServlet13() {
        super();
        // TODO Auto-generated constructor stub
    }
    /**
     * @see HttpServlet#doGet(HttpServletRequest request,HttpServletResponse response)
     */
    protected void doGet(HttpServletRequest request,HttpServletResponse response) throws ServletException,IOException {
        // TODO Auto-generated method stub
        response.setContentType("text/html;charset=utf-8");
        String data = "我热爱编程!";
        // 获取字符输出流对象
        PrintWriter print = response.getWriter();
        print.write(data);// 输出信息
    }
    /**
     * @see HttpServlet#doPost(HttpServletRequest request,HttpServletResponse response)
     */
    protected void doPost(HttpServletRequest request,HttpServletResponse response) throws ServletException,IOException {
```

```
        // TODO Auto-generated method stub
    }
}
```

运行结果如图 6-33 所示。

图 6-33 解决中文乱码问题的运行结果

2. 使用 HttpServletRequest 对象传递数据

我们学习了 HttpServletRequest 对象可以通过一系列方法获取客户端的各种请求信息，除此之外，我们还可以利用 HttpServletRequest 对象的属性操作方法实现数据的传递，其常用的属性操作方法如表 6-12 所示。

表 6-12 HttpServletRequest 对象常用的属性操作方法

方法	功能
setAttribute(String name, Object o)	第一个参数用于设置属性名，第二参数是属性值，该方法将 name 指定的属性与 Object 对象（即属性值）关联后存储进 ServletRequest 对象中
getAttribute(String name)	该方法用于从 HttpServletRequest 对象中返回指定名称的属性对象（即属性值）
removeAttribute(String name)	该方法用于从 HttpServletRequest 对象中删除指定名称的属性
getAttributeNames()	该方法用于返回一个包含 HttpServletRequest 对象中的所有属性名的 Enumeration 对象，在此基础上，可以对 ServletRequest 对象中的所有属性进行遍历处理

注意：只有属于同一个请求的数据才可以通过 HttpServletRequest 对象传递数据。

3. 请求转发

现在有一个验证用户登录信息的页面，我们希望用户在该页面提交用户名、密码后，首先验证用户名、密码是否正确，如果正确，则动态生成页面，显示"用户 XXX，欢迎您的登录！"注意该欢迎信息需要根据不同用户名动态生成，用户登录界面和登录成功界面分别如图 6-34、图 6-35 所示。

图 6-34 用户登录界面

图 6-35 登录成功界面

上述内容实现了两个功能，一个是验证用户名、密码，另一个是动态生成显示欢迎信息的页面，我们可以通过两个不同的 Servlet 程序实现这两个功能，当第一个 Servlet 程序验证用户名、密码正确后，就跳转到第二个 Servlet 程序，通过第二个 Servlet 程序动态生成欢迎信息。

这里的跳转是在服务器端实现的，最终返回客户端的响应结果由第二个 Servlet 程序生成。由此，我们利用 Servlet 程序的跳转功能实现了在服务器端将某项工作任务按功能模块进行分开处理。以上过程就是请求转发，即当前 Web 资源接收到请求信息后，如果不想处理请求信息，可以跳转到其他 Web 资源，并将当前请求信息传递给其他 Web 资源，由其他 Web 资源对当前请求信息进行处理并将响应结果提交给客户端。

Servlet 程序的跳转功能要通过 RequestDispatcher 实例对象实现，HttpServletRequest 接口提供了 getRequestDispatcher() 方法获取 RequestDispatcher 实例对象，其语法格式如下。

教学视频

```
RequestDispatcher getRequestDispatcher(String path)
```

注意：参数 path 指明了要跳转到的 Servlet 程序的路径，必须以"/"开头，用于表示当前 Web 应用的根目录，例如：getRequestDispatcher("/TestServlet15")。除了可以是指向 Servlet 程序的路径，参数 path 也可以是指向其他 Web 资源的路径，如 WEB-INF 目录中的内容。

getRequestDispatcher() 方法可用于实现 Servlet 程序跳转，那么请求信息如何传递过去呢？对此 RequestDispatcher 接口提供了 forward() 方法将请求信息从一个 Servlet 程序传递给另一个 Servlet 程序（或其他 Web 资源）。forward() 方法的语法格式如下所示。

例 6-24 代码

```
forward(ServletRequest request, ServletResponse response)
```

注意：forward() 方法必须在将响应结果返回给客户端之前调用。为了更好地理解，我们通过例 6-24、例 6-25、例 6-26 实现登录信息的请求转发。

【例 6-24】编写登录界面源代码，可扫描侧方二维码查看。

【例 6-25】请求转发 Servlet 程序，参考代码可扫描侧方二维码查看。

例 6-25 代码

【例 6-26】动态生成欢迎信息的参考代码，可扫描侧方二维码查看。

注意：在图 6-35 中，地址栏中显示的仍然是 TestServlet14 的请求路径，但是浏览器却显示 TestServlet15 的输出内容。这是因为请求转发是发生在服务器内部的行为，从 TestServlet14 到 TestServlet15 发生了一次请求转发，在这一次请求转发中可以使用 request 属性进行数据共享。

例 6-26 代码

任务 3　应用 Commons-FileUpload 和 Filter 实现文件上传与下载

 任务演示

在动态网站的实际运行过程中，很多网站都会涉及文件的上传和下载，如图片的上传、下载等。此外，为了网络安全，我们会设置 Filter 对用户请求进行过滤，或对某些特定的 IP

地址进行拦截，同时还使用 Filter 对中文乱码进行统一处理。文件上传界面、文件上传成功界面、文件下载界面、文件下载提示界面、地址过滤结果分别如图 6-36～图 6-40 所示。

图 6-36　文件上传界面　　　　　　　　图 6-37　文件上传成功界面

图 6-38　文件下载界面　　　　　　　　图 6-39　文件下载提示界面

图 6-40　地址过滤结果

知识准备

1. 文件上传流程

实现文件上传的两个操作如下：首先，在 Web 项目的页面中添加上传输入项；其次，在 Servlet 程序中读取上传文件的数据，并保存到目标路径中。

（1）在 Web 项目的页面中添加上传输入项

在 Web 项目的页面中添加上传输入项是指实现将文件提交给服务器的功能。在实际应用中，我们一般通过表单的形式将文件提交给服务器，如例 6-27 所示。

【例 6-27】在页面中添加上传输入项。

```
<%@ page language="java" import="java.util.*" pageEncoding="utf-8"%>
<!DOCTYPE HTML PUBLIC "-//W3C//DTD HTML 4.01 Transitional//EN">
<html>
  <head>
    <title>文件上传页面</title>
  </head>
  <body>
  <form  action="/chapter06/TestServlet16"  method="post"enctype="multipart/form-data">
```

```
        <p>用户名:<input type="text" name="name" placeholder="请填写用户名" ></p>
        <p>上传文件:<input type="file" name="myfile"></p>
        <p><input type="submit" value="提交"><input type="reset" value="重置"></p>
    </form>
    </body>
</html>
```

注意:

① 在实现文件上传时,表单的 method 属性必须设置为 POST 方法,enctype 属性必须设置为"multipart/form-data"。

② 在表单中使用 JSP 的"<input type="file">"标签添加上传文件时,必须设置 input 上传输入项的 name 属性,否则浏览器将不会发送上传文件的数据。

(2) 在 Servlet 程序中读取上传的文件数据

对于客户端上传的文件数据,我们可以在服务器端的 Servlet 程序中使用 getInputStream()方法进行读取。但用户可能会同时上传多个文件,因此在 Servlet 程序中直接读取上传的文件数据非常麻烦。对此,Apache 提供了开源的 Commons-FileUpload 组件,用于方便、高效地实现文件上传。注意,在使用 Commons-FileUpload 组件时,需要导入 commons-fileupload.jar 和 commons-io.jar 两个包,这两个包可以在 Apache 官网下载。

2. Commons-FileUpload 组件

Commons-FileUpload 组件主要通过 FileItem 接口、DiskFileItemFactory 类和 ServletFileUpload 类实现对上传的文件数据的读取和完成文件的上传。

(1) FileItem 接口

FileItem 接口主要用于封装单个表单字段元素的数据,一个表单字段元素对应一个 FileItem 对象,若同时上传三个文件,则在 form 表单中有三个"<input type="file">"标签,且每一个都被封装在一个对应的 FileItem 对象中。为了从封装的 FileItem 对象中获取表单字段元素的相关信息,FileItem 接口定义了相关方法,如表 6-13 所示。

表 6-13 FileItem 接口定义的相关方法

方法	功能
boolean isFormField()	isFormField()方法用于判断 FileItem 对象封装的数据是普通文本表单字段,还是文件表单字段,如果是普通文本表单字段则返回 True,否则返回 False
String getName()	getName()方法用于获取文件上传字段中的文件名。如果 FileItem 对象对应的是普通文本表单字段,getName()方法将返回 null,否则只要浏览器将文件的字段信息传递给服务器,getName()方法就会返回一个字符串类型的结果,如 C:\Sunset.jpg
String getFieldName()	getFieldName()方法用于获取表单字段元素描述头的 name 属性值,即表单标签 name 属性的值。如"name=file1"中的"file1"
void write(File file)	write()方法用于将 FileItem 对象中的主体内容保存到某个指定的文件中。如果 FileItem 对象中的主体内容保存在某个临时文件中,那么在退出该方法后,临时文件可能被清除。另外,该方法也可将普通表单字段的内容写入一个文件中,但它主要用于将上传的文件内容保存到本地文件系统中
String getString()	getString()方法用于将 FileItem 对象中的数据流内容以字符串形式返回。它有两个重载的定义形式:①public String getString(),②public String getString(java.lang.String encoding)。前者使用默认的字符集编码将数据流内容转换成字符串,后者使用参数指定的字符集编码将数据流内容转换成字符串
String getContentType()	getContentType()方法用于获取上传文件的类型,即表单字段元素描述头属性 Content-Type 的值,如"image/jpeg"。如果 FileItem 对象对应的是普通表单字段,则该方法将返回 null

注意：FileItem 接口支持序列化操作。

（2）DiskFileItemFactory 类

DiskFileItemFactory 类用于将请求消息中的文件封装成单独的 FileItem 对象。如果上传的文件比较小，则直接保存在内存中；如果上传的文件比较大，则会以临时文件的形式保存在磁盘的临时文件夹中。默认情况下，不管文件保存在内存中还是磁盘临时文件夹中，文件存储的临界值都是 10240 字节。DiskFileItemFactory 类有两个构造方法，如表 6-14 所示。

表 6-14 DiskFileItemFactory 类的构造方法

方法	功能
DiskFileItemFactory()	采用默认临界值和系统临时文件夹构造文件项工厂对象
DiskFileItemFactory(int sizeThreshold,File repository)	采用参数指定临界值和系统临时文件夹构造文件项工厂对象，第一个参数 sizeThreshold 表示文件保存的临界值，第二个参数 repository 表示临时文件的存储路径

（3）ServletFileUpload 类

ServletFileUpload 类是处理上传文件的核心高级类，通过调用 parseRequest(HttpServletRequest)方法可以将表单提交的数据封装成一个 FileItem 对象，并以 List 列表的形式返回。ServletFileUpload 类有两种构造方法，如表 6-15 所示。

表 6-15 ServletFileUpload 类的构造方法

方法	功能
ServletFileUpload()	构造一个未初始化的 ServletFileUpload 对象
ServletFileUpload(FileItemFactory fileItemFactory)	根据参数指定的 FileItemFactory 对象创建一个 ServletFileUpload 对象

注意：在文件上传过程中，若使用第一种构造方法创建 ServletFileUpload 对象，则需要在解析请求之前调用 setFileItemFactory()方法设置 fileItemFactory 属性。

3．文件下载流程

实现文件下载比较简单，一般不需要使用第三方组件，而是直接使用 Servlet 类和输入/输出流实现，总体流程如下。

① 当单击网页中的"下载"超链接时，客户端将请求信息提交给服务器对应的 Servlet 程序。

② 在服务器的 Servlet 程序中，首先获取文件的下载地址，并根据文件的下载地址创建文件字节输入流，然后通过输入流读取要下载的文件内容，最后将读取的内容通过输出流写到目标文件中。

注意：与访问服务器文件不同的是，实现文件的下载不仅需要指定文件的路径，还需要在 HTTP 协议中设置两个响应头字段，用来指定接收程序处理数据内容的方式为下载，具体如下。

```
// 设定接收程序处理数据的方式
Content-Disposition:attachment;
// 设定实体内容的 MIME 类型
filename = Content-Type:application/x-msdownload
```

4. Filter

Filter 又称过滤器，它位于客户端和服务器的 Servlet 等处理程序之间，能够对请求（request 对象）和响应（response 对象）进行检查和修改，实现 IP 过滤、统一字符编码和安全控制等通用操作。

（1）Filter 接口与 Filter 生命周期

Filter 有三个接口，分别是 Filter、FilterConfig 和 FilterChain，其中 Filter 接口是编写过滤器必须实现的接口，该接口定义了三种方法（init()、doFilter()、destroy()），如表 6-16 所示。

表 6-16　Filter 接口定义的方法

方法	功能
init(FilterConfig filterConfig)	init()方法是 Filter 接口的初始化方法，在创建 Filter 实例对象后将调用 init()方法。该方法的参数 filterConfig 用于读取 Filter 接口的初始化参数
doFilter (ServletRequest request, ServletResponse response, FilterChain chain)	doFilter()方法用于完成实际的过滤操作，当客户的请求满足过滤规则时，Servlet 容器将调用过滤器的 doFilter()方法完成实际的过滤操作。doFilter()方法有多个参数，其中，参数 request 和参数 response 为 Web 服务器或 Filter 链中的上一个 Filter 对象传递过来的请求和响应对象；参数 chain 代表当前 Filter 链的对象
destroy()	destroy()该方法用于释放被 Filter 对象打开的资源，如关闭数据库和 I/O 流。destroy()方法在 Web 服务器释放 Filter 对象之前被调用

表 6-16 中的三种方法是管理 Filter 生命周期的方法。Filter 生命周期指的是一个 Filter 对象从创建到执行，再到销毁的过程，可分为创建、执行、销毁三个阶段。

① 创建阶段。在启动服务器的时候会创建 Filter 对象，并调用 init()方法完成初始化。需要注意的是，在一次完整的请求中，Filter 对象只创建一次，init()方法也只执行一次。

② 执行阶段。当客户端请求目标资源时，服务器筛选出符合映射条件的 Filter 对象，并按照类名的先后顺序依次执行不同 Filter 对象的 doFilter()方法，如 Filter01 比 Filter02 优先执行。

③ 销毁阶段。当关闭服务器时，服务器调用 destroy()方法销毁 Filter 对象。

（2）FilterConfig 接口与 Filter 映射

FilterConfig 接口用于封装 Filter 对象的配置信息，在 Filter 对象初始化时，服务器将 FilterConfig 接口的对象作为参数传递给 Filter 对象的初始化方法。FilterConfig 接口的方法如表 6-17 所示。

表 6-17　FilterConfig 接口的方法

方法	功能
String getFilterName()	返回 Filter 对象的名称
ServletContext getServletContext()	返回 FilterConfig 接口的对象中封装的 ServletContext 对象
String getInitParameter(String name)	返回名为 "name" 的初始化参数值
Enumeration getInitParameterNames()	返回 Filter 对象的所有初始化参数的枚举

上文讲了 FilterConfig 接口用于封装 Filter 对象的配置信息，那么 Filter 对象的配置信息要如何设置呢？我们可以使用@WebFilter 注解配置 Filter 对象的相关信息，@WebFilter 注解标注在 Filter 类上，其常用属性如表 6-18 所示。

表 6-18 @WebFilter 注解的常用属性

属性名	功能
filterName	指定过滤器的名称。默认是过滤器类的名称
urlPatterns	指定一组过滤器的 URL 匹配模式
value	该属性等价于 urlPatterns 属性。urlPatterns 属性和 value 属性不能同时使用
servletNames	指定过滤器应用于哪些 Servlet。取值是@WebServlet 注解中的 name 属性的值
dispatcherTypes	指定过滤器的转发模式。具体取值包括 ERROR、FORWARD、INCLUDE、REQUEST
initParams	指定过滤器的一组初始化参数

在@WebFilter 注解中使用上述属性配置相关信息的过程就是 Filter 映射，Filter 映射方式分为两种。

① 使用通配符 "*" 拦截用户的所有请求。如果想让过滤器拦截用户的所有请求，则可以使用通配符 "*" 实现，代码如下。

```
@WebFilter(filterName="Filter02",urlPatterns="/*")
```

② 拦截不同访问方式的请求。通过将@WebFilter 注解的 dispatcherTypes 属性设置为不同的值（INCLUDE、FORWARD、ERROR），可以实现对不同访问方式的请求进行拦截。例如："@WebFilter（filterName = "ForwardFilter", urlPatterns = "/forward.jsp", dispatcherTypes= DispatcherType. FORWARD）"，该@WebFilter 注解将 dispatcherTypes 属性值设置为 "FORWARD"，因此，该过滤器可以拦截所有通过 forward()方法转发到 forward.jsp 页面的请求。

（3）FilterChain 接口与 Filter 链

如果多个 Filter 程序都对同一个 URL 的请求进行拦截，那么这些 Filter 程序就组成一个 Filter 链。Filter 链使用 FilterChain 接口的对象表示，FilterChain 接口提供了 doFilter()方法，该方法的作用是让 Filter 链上的当前 Filter 程序放行，使请求进入下一个 Filter 程序，如果当前 Filter 程序是 Filter 链上的最后一个过滤器程序，则将请求提交给处理程序。反之，当服务器对客户端请求做出响应时，响应结果也会被 Filter 链上的各个 Filter 程序拦截，拦截顺序与之前相反，最终响应结果被发送给客户端。注意，Filter 链上的各 Filter 程序的执行顺序是按照 Filter 程序的类名的排列顺序执行的，例如，Filter01 比 Filter02 优先执行。

任务实施

任务要求：

① 导入 Commons-FileUpload 组件。
② 用户通过 Upload.jsp 向 TestServlet16 程序提交上传文件。
③ 编写 TestServlet16 程序，在该程序中使用 Commons-FileUpload 组件获取上传文件信息，并将其保存到指定位置，输出提示信息。
④ 编写过滤器程序 Filter01，实现对中文乱码的统一处理。
⑤ 创建下载页面 Download.jsp，编写用于下载的超链接。
⑥ 编写 TestServlet17 程序，实现文件的下载。
⑦ 编写过滤器程序 Filter02，实现 IP 地址过滤，禁止本机上传文件。

具体实现步骤如下：

第一步：导入 Commons-FileUpload 组件。

将 commons-fileupload-1.2.2.jar 和 commons-io-2.4.jar 这两个包粘贴到项目 chapter06 的 WEB-INFO 文件夹下的 lib 文件夹中，如图 6-41 所示。

教学视频
（文件上传）

图 6-41　将两个包粘贴到指定文件夹中

第二步：用户通过 Upload.jsp 向 TestServlet16 程序提交上传文件，如例 6-28 所示。

【例 6-28】编写 Upload.jsp，源代码可扫描侧方二维码查看。

第三步：编写 TestServlet16 程序，用于保存上传文件，如例 6-29 所示。

【例 6-29】编写 TestServlet16 程序。

例 6-28 代码

```java
import java.io.*;
import java.util.*;
import javax.servlet.*;
import javax.servlet.annotation.MultipartConfig;
import javax.servlet.annotation.WebServlet;
import javax.servlet.http.*;
import org.apache.commons.fileupload.*;
import org.apache.commons.fileupload.disk.DiskFileItemFactory;
import org.apache.commons.fileupload.servlet.ServletFileUpload;
/**
 * Servlet implementation class TestServlet16
 */
@WebServlet(name="TestServlet16",urlPatterns="/TestServlet16")
public class TestServlet16 extends HttpServlet {
    private static final long serialVersionUID = 1L;

    /**
     * @see HttpServlet#HttpServlet()
     */
```

```java
    public TestServlet16() {
        super();
        // TODO Auto-generated constructor stub
    }
    /**
     * @see HttpServlet#doGet(HttpServletRequest request,HttpServletResponse response)
     */
    protected void doGet(HttpServletRequest request,HttpServletResponse response) throws ServletException,IOException {
        // TODO Auto-generated method stub
    }
    /**
     * @see HttpServlet#doPost(HttpServletRequest request,HttpServletResponse response)
     */
    protected void doPost(HttpServletRequest request,HttpServletResponse response) throws ServletException,IOException {
        // TODO Auto-generated method stub
        try {
            //设置ContentType字段值
            //response.setContentType("text/html;charset=utf-8");
            //创建DiskFileItemFactory工厂对象
            DiskFileItemFactory factory = new DiskFileItemFactory();
            //设置文件缓存目录,如果该目录不存在,则新建一个
            File f = new File("D:\\TempFolder");
            if (!f.exists()) {
                f.mkdirs();
            }
            //设置文件的缓存路径
            factory.setRepository(f);
            //创建ServletFileUpload对象
            ServletFileUpload fileupload = new ServletFileUpload(factory);
            //设置字符编码
            fileupload.setHeaderEncoding("utf-8");
            //解析request,得到上传文件的FileItem对象
            List<FileItem> fileitems = fileupload.parseRequest(request);
            //获取字符流
            PrintWriter writer = response.getWriter();
            //遍历集合
            for (FileItem fileitem:fileitems) {
                // 判断是否为普通字段
                if (fileitem.isFormField()) {
                    // 获得字段名和字段值
                    String name = fileitem.getFieldName();
                    if(name.equals("name")){
                        //如果文件不为空,则将其保存在value中
                        if(!fileitem.getString().equals("")){
```

```java
                        String value = fileitem.getString("utf-8");
                        writer.print("上传者:" + value + "<br />");
                    }
                }
            } else {
                //获取上传的文件名
                String filename = fileitem.getName();
                //处理上传文件
                if(filename != null && !filename.equals("")){
                    writer.print("上传的文件名称是:" + filename + "<br />");
                    //截取文件名
                    filename = filename.substring(filename.lastIndexOf ("\\") + 1);
                    //文件名唯一
                    filename = UUID.randomUUID().toString() + "_" + filename;
                    //在服务器中创建同名文件
                    String webPath = "/upload/";
                    //将服务器中的文件夹路径与文件名组合成完整的服务器路径
                    String filepath = getServletContext()
                            .getRealPath(webPath + filename);
                    //创建文件
                    File file = new File(filepath);
                    file.getParentFile().mkdirs();
                    file.createNewFile();
                    //获得上传文件流
                    InputStream in = fileitem.getInputStream();
                    //使用 FileOutputStream 打开服务器的上传文件
                    FileOutputStream out = new FileOutputStream(file);
                    //流的拷贝
                    byte[] buffer = new byte[1024];//每次读取1字节
                    int len;
                    //开始读取上传文件,并将其输出到服务器的上传文件输出流中
                    while ((len = in.read(buffer)) > 0)
                        out.write(buffer,0,len);
                    //关闭流
                    in.close();
                    out.close();
                    //删除临时文件
                    fileitem.delete();
                    writer.print("upload success!<br />");
                }
            }
        }
    } catch (Exception e) {
        throw new RuntimeException(e);}
    }
}
```

第四步：运行 Upload.jsp 实现文件上传，并查看文件上传结果。

① 启动 Tomcat 服务器，在浏览器地址栏输入访问地址，得到如图 6-42 所示的文件上传界面。

② 在图 6-42 的文件上传界面中单击"浏览"按钮，选择要上传的文件，添加上传文件如图 6-43 所示。

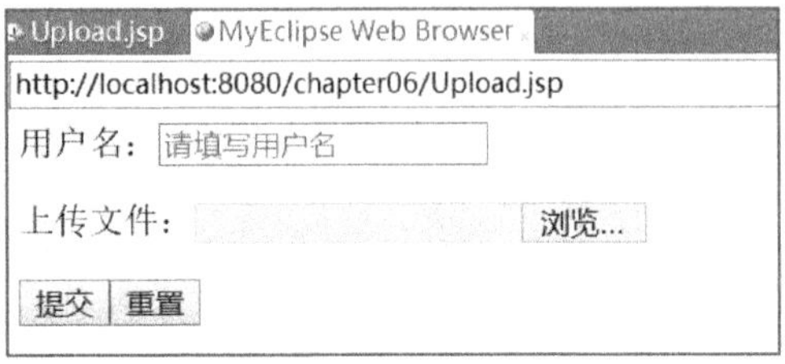

图 6-42　文件上传界面 　　　　　图 6-43　添加上传文件

③ 在图 6-43 中单击"提交"按钮实现文件上传，文件上传成功界面如图 6-44 所示。

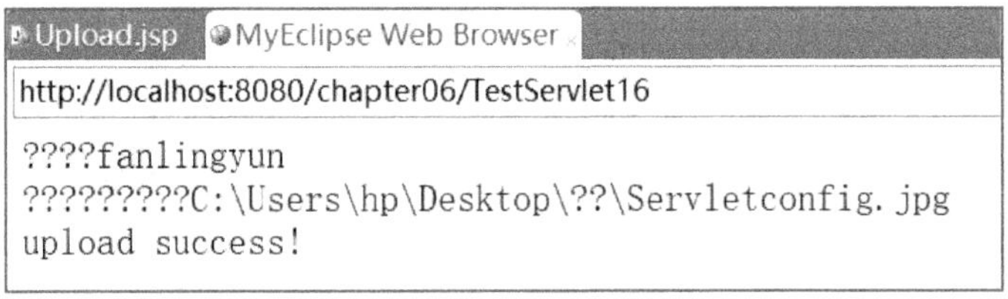

图 6-44　文件上传成功界面

在图 6-44 中，虽然提示上传成功，但是所有的中文字符都是乱码，因此，我们需要进一步设计过滤器程序，统一对所有中文乱码问题进行处理。

第五步：创建过滤器程序 Filter01，用于统一处理中文乱码问题。

① 右键单击 chapter06 中的 src 文件夹，依次选择"New"→"Filter"选项，如图 6-45 所示。

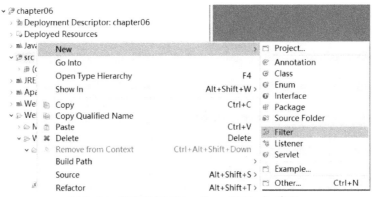

图 6-45　依次选择"New"→"Filter"选项

② 在"Create Filter"对话框中输入"Class name"的值为"Filter01"，如图 6-46 所示。

③ 在图 6-46 中单击"Finish"按钮新建过滤器程序 Filter01，过滤器程序 Filter01 的代码如例 6-30 所示。

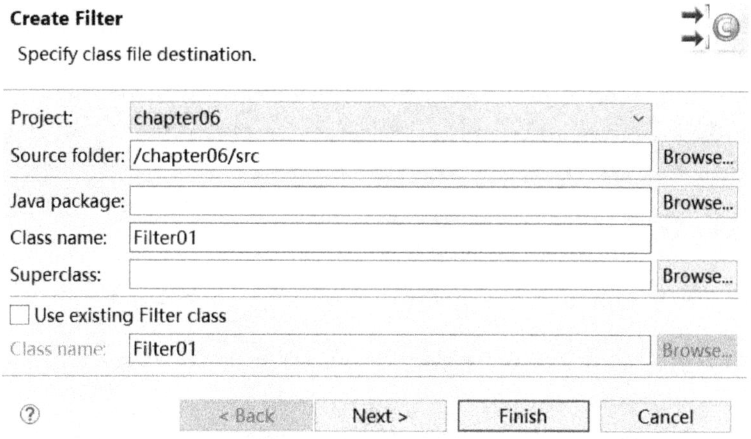

图 6-46 "Create Filter"对话框

【例 6-30】编写过滤器程序 Filter01 的代码。

```java
import java.io.IOException;
import javax.servlet.Filter;
import javax.servlet.FilterChain;
import javax.servlet.FilterConfig;
import javax.servlet.ServletException;
import javax.servlet.ServletRequest;
import javax.servlet.ServletResponse;
import javax.servlet.annotation.WebFilter;
/**
 * Servlet Filter implementation class Filter01
 */
@WebFilter("/Filter01")
public class Filter01 implements Filter {
    /**
     * Default constructor.
     */
    public Filter01() {
        // TODO Auto-generated constructor stub
    }
    /**
     * @see Filter#destroy()
     */
    public void destroy() {
        // TODO Auto-generated method stub
    }
    /**
     * @see Filter#doFilter(ServletRequest,ServletResponse,FilterChain)
     */
    public   void   doFilter(ServletRequest   request,ServletResponse   response,
```

```
FilterChain chain) throws IOException,ServletException {
        // TODO Auto-generated method stub
        // place your code here
        // pass the request along the filter chain
        chain.doFilter(request,response);
    }
    /**
     * @see Filter#init(FilterConfig)
     */
    public void init(FilterConfig fConfig) throws ServletException {
        // TODO Auto-generated method stub
    }
}
```

④ 修改例 6-30 中的过滤器程序 Filter01 的代码，如例 6-31 所示。

【例 6-31】修改后的过滤器程序 Filter01 的代码如下所示。

```
import java.io.IOException;
import javax.servlet.Filter;
import javax.servlet.FilterChain;
import javax.servlet.FilterConfig;
import javax.servlet.ServletException;
import javax.servlet.ServletRequest;
import javax.servlet.ServletResponse;
import javax.servlet.annotation.WebFilter;
/**
 * Servlet Filter implementation class Filter01
 */
@WebFilter(filterName="Filter01",urlPatterns="/*")
public class Filter01 implements Filter {
    /**
     * Default constructor.
     */
    public Filter01() {
        // TODO Auto-generated constructor stub
    }
    /**
     * @see Filter#destroy()
     */
    public void destroy() {
        // TODO Auto-generated method stub
    }
    /**
     * @see Filter#doFilter(ServletRequest,ServletResponse,FilterChain)
     */
    public void doFilter(ServletRequest request,ServletResponse response,
FilterChain chain) throws IOException,ServletException {
```

```
            // TODO Auto-generated method stub
            //设置请求中的字符编码为UTF-8
            request.setCharacterEncoding("utf-8");
            //设置相应的字符编码为UTF-8
            response.setContentType("text/html;charset=utf-8");
            //调用下一个过滤器
            chain.doFilter(request,response);
        }
        /**
         * @see Filter#init(FilterConfig)
         */
        public void init(FilterConfig fConfig) throws ServletException {
            // TODO Auto-generated method stub
        }
    }
```

⑤ 再次运行 upload.jsp 实现文件上传，发现中文乱码问题已经解决，运行结果如图 6-47 所示。

图 6-47　运行结果

第六步：创建下载页面 Download.jsp，编写用于下载的超链接。

① 右键单击 WebRoot 文件夹，依次选择"New"→"Folder"选项，新建文件夹菜单如图 6-48 所示。

② 在"Folder"对话框中输入"Folder name"的值为"download"，如图 6-49 所示。

③ 在图 6-49 中单击"Finish"按钮，得到新建文件夹 download 如图 6-50 所示。

④ 拷贝图片 Servletconfig.jpg 到新建文件夹 download 中，如图 6-51 所示。

教学视频
（文件下载）

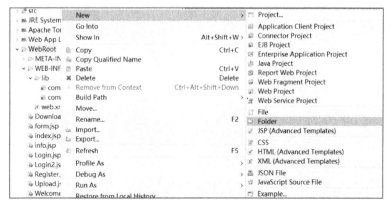

图 6-48　新建文件夹菜单

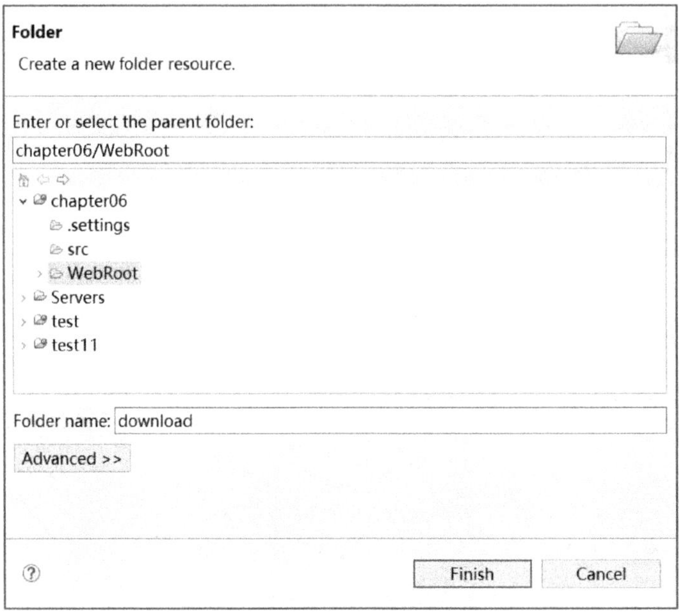

图 6-49 "Folder"对话框

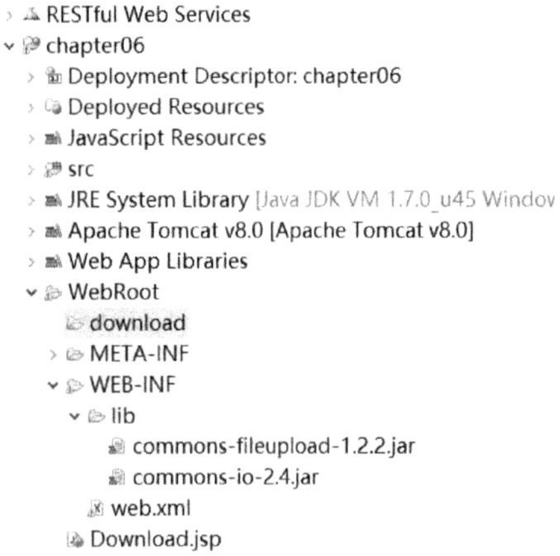

图 6-50 得到新建文件夹 download

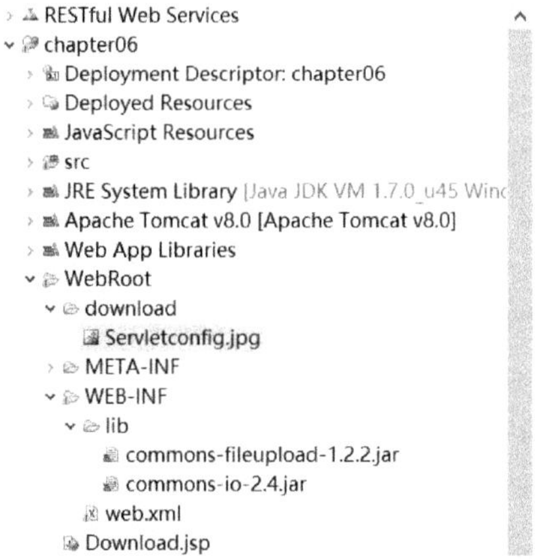

图 6-51 拷贝图片 Servletconfig.jpg 到新建文件夹 download 中

⑤ 创建下载页面 Download.jsp 如例 6-32 所示。

【例 6-32】编写下载页面 Download.jsp 的源代码，可扫描侧方二维码查看。

第七步：编写 TestServlet17 程序，实现文件下载，如例 6-33 所示。

【例 6-33】编写 TestServlet17 程序的源代码。

例 6-32 代码

```
import java.io.IOException;
import javax.servlet.ServletException;
import javax.servlet.annotation.WebServlet;
import javax.servlet.http.HttpServlet;
import javax.servlet.http.HttpServletRequest;
import javax.servlet.http.HttpServletResponse;
```

```java
import java.io.*;
/**
 * Servlet implementation class TestServlet17
 */
@WebServlet("/TestServlet17")
public class TestServlet17 extends HttpServlet {
    private static final long serialVersionUID = 1L;

    /**
     * @see HttpServlet#HttpServlet()
     */
    public TestServlet17() {
        super();
        // TODO Auto-generated constructor stub
    }
    /**
     * @see HttpServlet#doGet(HttpServletRequest request,HttpServletResponse response)
     */
    protected void doGet(HttpServletRequest request,HttpServletResponse response) throws ServletException,IOException {
        // TODO Auto-generated method stub
        //获取要下载的文件名称
        String filename = request.getParameter("filename");
        //下载文件所在目录
        String folder = "/download/";
        // 通知浏览器以下载的方式打开
        response.addHeader("Content-Type","application/octet-stream");
        response.addHeader("Content-Disposition",
                "attachment;filename="+filename);
        folder=folder+filename;
        // 通过文件流读取文件
        InputStream in = getServletContext().getResourceAsStream(folder);
        // 获取 response 对象的输出流
        OutputStream out = response.getOutputStream();
        byte[] buffer = new byte[1024];
        int len;
        //循环取出文件流中的数据
        while ((len = in.read(buffer)) != -1) {
            out.write(buffer,0,len);
        }
    }
    /**
     * @see HttpServlet#doPost(HttpServletRequest request,HttpServletResponse response)
     */
    protected void doPost(HttpServletRequest request,HttpServletResponse response) throws ServletException,IOException {
        // TODO Auto-generated method stub
```

        }
}

第八步：运行下载页面 Download.jsp，得到下载页面如图 6-52 所示，单击图 6-52 中的"文件下载"超链接，可打开如图 6-53 所示的文件下载窗口。在图 6-53 中单击"保存"按钮完成文件下载。

图 6-52　下载页面

图 6-53　文件下载窗口

第九步：编写过滤器程序 Filter02，实现 IP 地址过滤，如例 6-34 所示。

【例 6-34】编写过滤器程序 Filter02 的源代码。

```
import java.io.IOException;
import java.io.PrintWriter;
import javax.servlet.Filter;
import javax.servlet.FilterChain;
import javax.servlet.FilterConfig;
import javax.servlet.ServletException;
import javax.servlet.ServletRequest;
import javax.servlet.ServletResponse;
import javax.servlet.annotation.WebFilter;
/**
 * Servlet Filter implementation class Filter02
 */
@WebFilter(filterName="Filter02",urlPatterns="/*")
public class Filter02 implements Filter {
    /**
     * Default constructor.
     */
    public Filter02() {
        // TODO Auto-generated constructor stub
    }
    /**
```

```java
     * @see Filter#destroy()
     */
    public void destroy() {
      // TODO Auto-generated method stub
    }
    /**
     * @see Filter#doFilter(ServletRequest,ServletResponse,FilterChain)
     */
    public void doFilter(ServletRequest request,ServletResponse response,
FilterChain chain) throws IOException,ServletException {
      // TODO Auto-generated method stub
      //获取请求信息的IP地址
      String ip=request.getRemoteAddr();
      //判断IP地址是否符合要求,如果是本机地址,则拦截请求,不允许本机上传文件,不执行
FilterChain的doFilter()方法
      if(ip.equals("0:0:0:0:0:0:0:1")){//0:0:0:0:0:0:0:1是IPv6地址,类同IPV4地
址127.0.0.1
        //输出得到的编码信息
        PrintWriter out=response.getWriter();
        out.print("IP地址非法!");
      }else{//IP地址合法,请求继续
        chain.doFilter(request,response);
      }
    }
    /**
     * @see Filter#init(FilterConfig)
     */
    public void init(FilterConfig fConfig) throws ServletException {
      // TODO Auto-generated method stub
    }
  }
```

再次运行 Upload.jsp,此时过滤器程序 Filter02 会对本机 IP 进行拦截,效果如图 6-54 所示。

图 6-54 对本机 IP 进行拦截的效果

任务拓展

1. 认识 Listener

在动态网站开发中,经常需要对 session、request、context 等事件进行监听,以便及时做出处理。为此,Servlet 提供了监听器(Listener),Listener 是一个实现了特定接口的 Java 程

序，专门用于监听 Web 应用程序中 ServletContext、HttpSession 和 ServletRequest 等域对象的创建和销毁过程，以及这些域对象属性的修改过程，并且能够感知绑定到 HttpSession 域中的某个对象的状态变化。Listener 共有 8 个不同的监听器接口，如表 6-19 所示。

表 6-19 Listener 的监听器接口

监听器接口	描述
ServletContextListener	用于监听 ServletContext 对象的创建与销毁过程
HttpSessionListener	用于监听 HttpSession 对象的创建和销毁过程
ServletRequestListener	用于监听 ServletRequest 对象的创建和销毁过程
ServletContextAttributeListener	用于监听 ServletContext 对象中的属性变更过程
HttpSessionAttributeListener	用于监听 HttpSession 对象中的属性变更过程
ServletRequestAttributeListener	用于监听 ServletRequest 对象中的属性变更过程
HttpSessionBindingListener	用于监听将 JavaBean 对象绑定到 HttpSession 对象和 HttpSession 对象解绑的事件
HttpSessionActivationListener	用于监听 HttpSession 对象活化和钝化的过程

根据监听事件的不同，可以将监听器接口分为三类。

① 监听 ServletContext、HttpSession 和 ServletRequest 的创建与销毁过程：ServletContextListener、HttpSessionListener、ServletRequestListener。

② 监听 ServletContext、HttpSession 和 ServletRequest 对象的属性变化过程：ServletContextAttributeListener、HttpSessionAttributeListener、ServletRequestAttributeListener。

③ 监听绑定到 HttpSession 的对象的状态变化过程：HttpSessionBindingListener、HttpSessionActivationListener。

2. 创建 Listener

在 Servlet 规范中，监听器都定义了相应的接口，在编写监听器程序时只需实现对应的接口就可以。接下来，我们以 HttpSessionListener 监听器接口的实现为例，演示如何创建 HttpSessionListener 监听器，并监听 HttpSession 对象的创建与销毁过程。

① 新建测试页面 testServletRequestListener.jsp，如例 6-35 所示。

【例 6-35】编写测试页面 testServletRequestListener.jsp 的源代码。

```
<%@ page language="java" import="java.util.*" pageEncoding="utf-8"%>
<!DOCTYPE HTML PUBLIC "-//W3C//DTD HTML 4.01 Transitional//EN">
<html>
  <head>
    <title>ServletRequestListener 监听器接口</title>
  </head>
  <body>
      测试 ServletRequestListener 监听器接口。
  </body>
</html>
```

② 右键单击 src 文件夹，依次选择"New"→"Listener"选项，如图 6-55 所示。

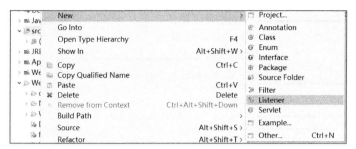

图 6-55 依次选择"New"→"Listener"选项

③ 先在"Specify class file destination"对话框中，输入"Class name"的值为"TestListener"，如图 6-56 所示，然后单击"Next"按钮，进入如图 6-57 所示的"Select the application lifecycle events to listen to"对话框。

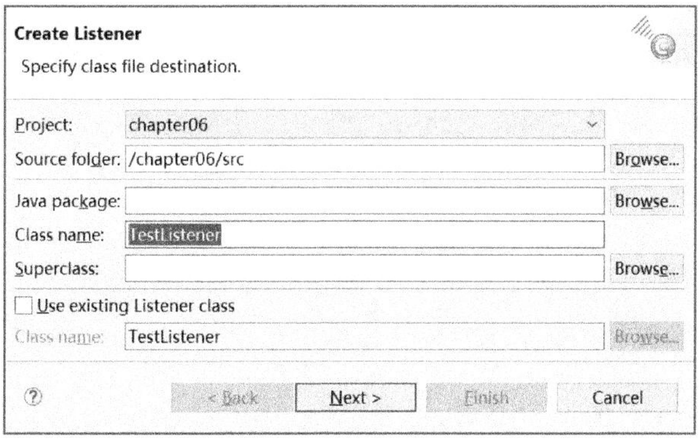

图 6-56 "Specify class file destination"对话框

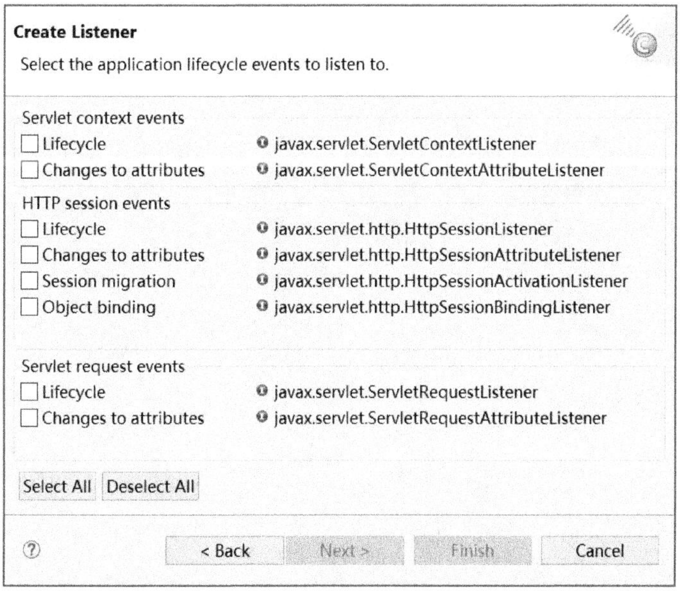

图 6-57 "Select the application lifecycle events to listen to"对话框

④ 先在图 6-57 中选择"Servlet request events"下方的"LifeCycle"选项，如图 6-58 所示，然后单击"Finish"按钮，最后生成如例 6-36 所示的 TestListener.java 源代码。

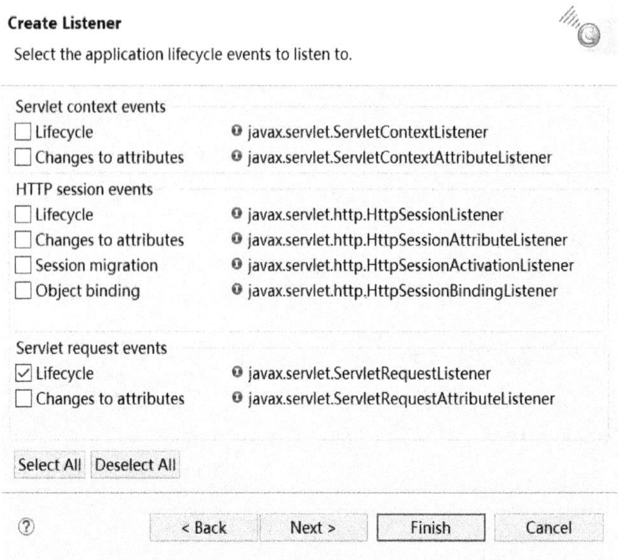

图 6-58 选中"Servlet request events"下面的"LifeCycle"选项

【例 6-36】TestListener.java 源代码。

```java
import javax.servlet.ServletRequestEvent;
import javax.servlet.ServletRequestListener;
import javax.servlet.annotation.WebListener;
/**
 * Application Lifecycle Listener implementation class TestListener
 *
 */
@WebListener
public class TestListener implements ServletRequestListener {
    /**
     * Default constructor.
     */
    public TestListener() {
        // TODO Auto-generated constructor stub
    }
    /**
     * @see ServletRequestListener#requestDestroyed(ServletRequestEvent)
     */
    public void requestDestroyed(ServletRequestEvent arg0)  {
        // TODO Auto-generated method stub
    }
    /**
     * @see ServletRequestListener#requestInitialized(ServletRequestEvent)
     */
    public void requestInitialized(ServletRequestEvent arg0)  {
        // TODO Auto-generated method stub
    }
}
```

注意：在上述代码中的@WebListener 注解处开启了 Listener。

⑤ 在例 6-36 的 requestInitialized()方法和 requestDestroyed()方法中分别添加对 ServletRequest 对象创建和销毁事件的处理代码，如例 6-37 所示。

【例 6-37】修改后的 TestListener.java 源代码如下。

```java
import javax.servlet.ServletRequestEvent;
import javax.servlet.ServletRequestListener;
import javax.servlet.annotation.WebListener;
/**
 * Application Lifecycle Listener implementation class TestListener
 *
 */
@WebListener
public class TestListener implements ServletRequestListener {
    /**
     * Default constructor.
     */
    public TestListener() {
        // TODO Auto-generated constructor stub
    }
    /**
     * @see ServletRequestListener#requestDestroyed(ServletRequestEvent)
     */
    public void requestDestroyed(ServletRequestEvent arg0) {
        // TODO Auto-generated method stub
        System.out.println("ServletRequest 对象被销毁了!");
    }
    /**
     * @see ServletRequestListener#requestInitialized(ServletRequestEvent)
     */
    public void requestInitialized(ServletRequestEvent arg0) {
        // TODO Auto-generated method stub
        System.out.println("ServletRequest 对象被创建成功!");
    }
}
```

⑥ 运行测试页面 testHttpSessionListener.jsp，在控制台输出如图 6-59 所示的 ServletRequest 对象的创建与销毁信息。

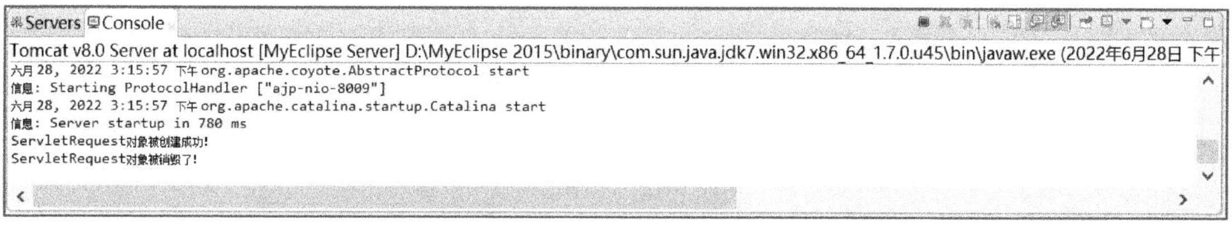

图 6-59 ServletRequest 对象的创建与销毁信息

项目实训

实训一 统计站点的访问次数

要求：编写 Servlet 程序，利用 ServletContext 对象统计站点的访问次数，每刷新一次页面，访问次数增加 1，站点的访问如图 6-60 所示。

图 6-60 站点的访问

实训二 设计用户注册界面

要求：在用户注册界面输入用户信息，提交后显示用户信息（应用 Servlet 中的 HttpRequest、HttpResponse 创建对象），提交用户信息界面、显示用户信息界面分别如图 6-61、图 6-62 所示。

图 6-61 提交用户信息界面

图 6-62 显示用户信息界面

实训三 设计过滤器

要求：过滤以字母"F"开头的用户名，使其重定向到 error.jsp 页面，其他用户名直接登录到 welcome.jsp 页面，用户登录界面、信息输入错误提示界面、信息输入正确提示界面分别如图 6-63、图 6-64、图 6-65 所示。

图 6-63　用户登录界面

图 6-64　信息输入错误提示界面

图 6-65　信息输入正确提示界面

课 后 练 习

一、填空题

1. Servlet 的生命周期分为三个阶段：_____阶段、_____阶段和_____阶段。
2. 假设 form 表单的提交方式为 POST，那么在 Servlet 中调用的是_____方法。
3. 通过@WebFilter 注解声明一个过滤器对象。在此注解中包含两个常用属性，分别为_____、_____。

二、选择题

1. 若要针对 HTTP 请求撰写 Servlet 类别，以下做法正确的是（　　）。
   A. 继承 Servlet 接口　　　　　　　　B. 继承 GenericServlet
   C. 继承 HttpServlet　　　　　　　　D. 直接定义一个结尾名称为 Servlet 的类别
2. 针对 HTTP 的 GET 请求进行处理与响应的是（　　）。
   A. 重新定义 service()方法　　　　　B. 重新定义 doGet()方法
   C. 定义一个 doService()方法　　　　D. 定义一个 get()方法
3. 在@WebServlet 注解中定义了以下内容：

```
@WebServlet(name="GoodBye",urlPatterns="/goodbye")
```

可以正确要求 Servlet 进行请求处理的 URL 是（　　）。

A. /GoodBye　　　B. /goodbye.do　　　C. /LoguotServlet　　　D. /goodbye

4. 在 Servlet 容器中，以下（　　）两个类别的实例分别代表 HTTP 的请求对象与响应对象。

A. HttpRequest　　　　　　　　　　　B. HttpServletRequest

C. HttpServletResponse　　　　　　　D. HttpPrintWriter

5. 可以取得 password 请求参数的值的程序代码是（　　）。

A. request.getParameter("password");

B. request.getParameters("password")[0];

C. request.getParameterValues("password")[0];

D. request.getRequestParameter("password");

6. 关于过滤器的描述正确的是（　　）。

A. Filter 接口定义了 init()、service()与 destroy()方法

B. 会传入 ServletRequest 与 ServletResponse，最后传至 Filter

C. 要执行下一个过滤器，必须先执行 FilterChain 的 next()方法

D. 如果要取得初始参数，则要使用 FilterConfig 对象

7. 关于 FilterChain 的描述正确的是（　　）。

A. 如果不呼叫 FilterChain 的 doFilter()方法，则请求略过下一个过滤器，直接交给 Servlet

B. 如果有下一个过滤器，则在呼叫 FilterChain 的 doFilter()方法时，将请求交给下一个过滤器

C. 如果没有下一个过滤器，则在呼叫 FilterChain 的 doFilter()方法时，将请求交给 Servlet

D. 如果没有下一个过滤器，则呼叫 FilterChain 的 doFilter()方法没有作用

项目七　EL 和 JSTL 技术

☀ 项目要求

本项目是 EL 和 JSTL 技术的应用，要求完成 EL 表达式及 JSTL 的应用，访问 Web 作用域对象中的数据。

☀ 项目分析

要完成项目任务，至少需要具备两个基本条件：一是掌握常用的 EL 运算符和 EL 函数，二是掌握 JSTL 的基本用法和核心标签库的使用方法。该项目分为 2 个任务，分别是应用 JSTL 实现用户个人信息获取、应用 JSTL 实现商品展示。

☀ 项目目标

【知识目标】熟悉 EL 和 JSTL 的含义，掌握 EL 运算符和 EL 函数，掌握 JSTL 的基本应用，掌握自定义标签库的开发技术。

【能力目标】能够用 EL 表达式访问隐式对象，能使用 EL 函数，能应用自定义标签库的开发技术。

【素质目标】提高学生发现问题、分析问题、解决问题的能力。

知识导图

```
                        ┌─ EL表达式
                        ├─ EL表达式语法
                        ├─ EL表达式标识符
                        ├─ EL关键字
          ┌─ 任务1 应用JSTL实现 ─┼─ EL变量与常量
          │  用户个人信息获取    ├─ EL访问数据
          │                    ├─ EL运算符
          │                    ├─ EL运算符的优先级
          │                    ├─ 使用EL表达式从作用域中获取数据
          │                    ├─ EL的隐式对象
          │                    └─ 任务拓展 ─┬─ 应用EL获取Cookie对象的信息
EL和JSTL技术                               └─ 应用EL获取initParam对象的信息
          │                    ┌─ JSTL的概念
          │                    ├─ 下载和安装JSTL
          └─ 任务2 应用JSTL实现 ─┤                       ┌─ 表达式标签
             商品展示           ├─ JSTL的核心标签库 ─────┼─ 流程控制标签
                                │                       └─ 循环标签
                                │                       ┌─ <c:param>标签
                                └─ 任务拓展 ────────────┼─ <c:redirect>标签
                                                        └─ <c:url>标签
```

任务 1　应用 JSTL 实现用户个人信息获取

任务演示

在客户端有一份用户信息表,需要用户填写信息并提交,应用 EL 表达式将用户信息显示在页面上。用户信息提交界面、获取用户信息界面分别如图 7-1、图 7-2 所示。

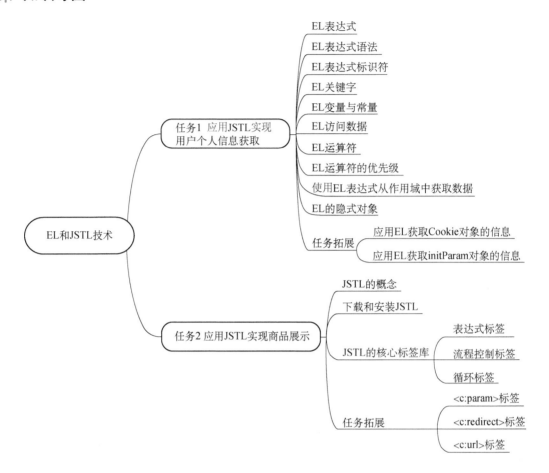

图 7-1　用户信息提交界面

图 7-2 获取用户信息界面

知识准备

1. EL 表达式

EL（Express Language）表达式是 JSP2.0 引入的新内容，可以简化 JSP 开发中的对象引用，规范页面代码，增强程序的可读性和可维护性。EL 表达式具有以下特点。

① EL 表达式可以与 JavaScript 语句结合使用。
② EL 表达式可以自动进行类型转换。
③ EL 表达式不仅可以访问一般变量，还可以访问 JavaBean 中的属性、嵌套属性和集合对象。
④ EL 表达式可以执行算术运算、逻辑运算、关系运算和条件运算等。
⑤ EL 表达式可以获取 pageContext 对象，进而获取其他内置对象。
⑥ EL 表达式可以访问 JSP 的作用域（如 request、session、application 和 page）。

2. EL 表达式语法

EL 表达式以"$"开头，表达式的内容包含在"{}"中，具体格式如下。

```
${表达式}
```

上式中的表达式必须符合 EL 表达式的语法要求。

【多学一招】
要在 JSP 网页中显示"${}"字符串有两种方法，第一种是在字符串前面加上"\"，即"\${}"；第二种是嵌套单引号，即"${'${'}}"。

【例 7-1】应用 EL 表达式从 Servlet 中获取用户名和密码。
第一步：创建一个 MyServletel。

教学视频

```
package el.com;
import java.io.IOException;
import javax.servlet.RequestDispatcher;
import javax.servlet.ServletException;
import javax.servlet.annotation.WebServlet;
import javax.servlet.http.HttpServlet;
import javax.servlet.http.HttpServletRequest;
import javax.servlet.http.HttpServletResponse;
@WebServlet("/MyServletel")
public class MyServletel extends HttpServlet {
    private static final long serialVersionUID = 1L;
```

```java
        protected void doGet(HttpServletRequest request,HttpServletResponse response)
throws ServletException,IOException {
        request.setAttribute("username","lili");
        request.setAttribute("password","123");
        RequestDispatcher dispatcher=request.getRequestDispatcher("/El1.jsp");
        dispatcher.forward(request,response);
    }
        protected void doPost(HttpServletRequest request,HttpServletResponse response)
throws ServletException,IOException {
        doGet(request,response);
    }
}
```

第二步：创建一个 El1.jsp。

```jsp
<%@ page language="java" contentType="text/html;charset=utf-8"
    pageEncoding="utf-8"%>
<!DOCTYPE html>
<html>
<head>
<meta charset="utf-8">
<title>Insert title here</title>
</head>
<body>
用户名:<%=request.getAttribute("username") %> <br/>
密码:<%=request.getAttribute("password") %> <br/>
<hr>
<b>使用 EL:</b></br>
用户名:${username} <br/>
密码:${password}<br/>
</body>
</html>
```

运行结果如图 7-3 所示。

图 7-3 应用 EL 表达式从 Servlet 中获取用户名和密码的运行结果

我们使用 EL 表达式成功获取了 Servlet 中的数据，可以看出 EL 表达式明显简化了 JSP 页面的书写，使程序简洁且易维护。另外，当域对象里面的值不存在时，使用 EL 表达式获取域对象里面的值会返回空字符串；而使用 Java 方式获取时，如果返回值是 null，则会报空指针异常，所以在现实开发中推荐使用 EL 表达式获取域对象中存储的数据。

3. EL 表达式标识符

在 EL 表达式中,大部分变量、函数的名称都是由程序员自定义的,我们把这些名称称为标识符。标识符除了要遵循最基本的命名规范,还有一些必须要遵守的要求。

① 不能以数字开头。
② 不能是 EL 表达式中的隐式对象。
③ 不能是 EL 表达式中的关键字。
④ 不能包含单引号(')、双引号(")、斜线(/)等特殊符号。

比如:username、userpassword、_name 都是合法的标识符,而 12name、user-name、user'pass 都是不合法的标识符。

4. EL 关键字

关键字是编程语言里事先定义好并被赋予了特殊含义的单词,EL 中定义了许多关键字。EL 中常用的关键字如表 7-1 所示。

表 7-1 EL 中常用的关键字

关键字				
and	eq	gt	True	instanceof
div	or	ne	le	False
lt	null	mod	not	ge

5. EL 变量与常量

(1)变量

EL 中的变量是一个基本的存储单元,不用事先定义就可以直接使用。EL 可以将变量映射到对象上。

【例 7-2】从后端 Servlet 向前端 JSP 页面传入"重庆城市职业学院"字符串,并使用 EL 变量展示出来,展示效果如图 7-4 所示。

教学视频

后端 Servlet:

```
request.setAttribute("school","重庆城市职业学院");
```

前端 JSP 页面:

```
${school}
```

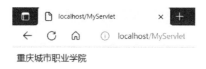

图 7-4 EL 变量的展示效果

这里的"school"就是 EL 变量,可直接访问 school 的值。依次到 pageContext 中找 school,如果没有则到 request 作用域中找,再没有就到 session 作用域中找,还没有就到 application Context 中找,再没有找到就是真的没有了。

(2)常量

常量不能被改变,EL 包含多种常量,下面一一介绍。

① 整型常量

整型常量与 Java 语言中的十进制整型常量相同，取值范围是 $(-2)^{63} \sim 2^{63}$。

② 浮点型常量

浮点型常量可以用整数部分加小数部分表示，也可以用指数形式表示，取值范围与 Java 语言中 double 相同，即 4.9E-324～1.8E308。

③ 布尔型常量

布尔型常量与 Java 语言中的 boolean 相同，取值分别为 True 和 False。

④ 字符串型常量

字符串型常量是用单引号和双引号引起来的一串字符。需要注意的是，如果需要表示字符的本身就是单引号或双引号，则需要使用转义字符进行转义。这里的转义字符就是反斜杠（\），如 "\'" 或 "\""。

⑤ null 常量

null 常量表示引用的对象是空的，它只有一个 null 值。

6. EL 访问数据

正常情况下，在 EL 中写出 Servlet 传入的变量名称就可直接访问数据，但如果从 Servlet 传入的是一个复杂类型的数据，如 JavaBean 对象、集合对象，就需要用到 "." 或 "[]" 来访问。

（1）点运算符（.）

使用点运算符可直接访问到对象中的属性或数组中的某个元素，假如 Servlet 传入一个名为 student 的对象，而这个对象中有一个 name 属性，则访问这个 name 属性的方法如下。

```
${student.name}
```

（2）方括号运算符（[]）

与点运算符一样，方括号运算符也可以访问对象中的属性或数组中的某个元素。不同之处在于，若表达式中包含了一些特殊符号，则必须用方括号运算符访问，方法如下。

```
${student["user-name"]}
```

【脚下留心】

方括号运算符还可以访问 List 集合或数组中的某个元素，如${students[1]}，其中数字 1 是集合或数组中的索引值，表示访问的是 students 集合中第二个元素。同时，方括号运算符和点运算符可以混合使用，如${students[1].username}。

7. EL 运算符

大部分 EL 运算符和 Java 运算符的功能相同。常见的 EL 运算符可分为以下几种。

（1）算术运算符

常见的算术运算符如表 7-2 所示。

表 7-2 常见的算术运算符

运算符	表达式
+（加）	${1+2}

(续表)

运算符	表达式
-（减）	${1-2}
*（乘）	${1*2}
/或 div（除）	${1/2}或${1 div 2}
%或 mod（取模）	${1%2}或${1 mod 2}

【脚下留心】
在做加法运算时，一定要注意 EL 运算符和 Java 运算符的不同之处。在 Java 中，字符串与数字相加，最终结果为字符串，如"1"+2 结果为"12"，"1a"+2 结果为"1a2"。而在 EL 中使用两种不同数据类型的数据进行运算时，会先尝试对数据类型进行转换，若能够转换，则正常进行运算；若不能够转换，则会报错。例如，${"1"+2}结果为3，而${"1a"+2}的结果会报错。

（2）比较运算符

常见的比较运算符如表 7-3 所示。

表 7-3 常见的比较运算符

运算符	表达式
>或 gt（大于）	${1>1}或${1 gt 1}
<或 lt（小于）	${1<1}或${1 lt 1}
==或 eq（等于）	${1==1}或${1 eq 1}
!=或 ne（不等于）	${1!=1}或${1 ne 1}
>=或 ge（大于等于）	${1>=1}或${1 ge 1}
<=或 le（小于等于）	${1<=1}或${1 le 1}

说明：因为 JSP 页面不可避免地会用到前端页面标签，而标签需要用"< >"括起来，所以经常出现页面标签中的"< >"与大于符号、小于符号产生冲突。建议使用"gt""lt"英文缩写来表示大于符号或小于符号。

（3）逻辑运算符

常见的逻辑运算符如表 7-4 所示。

表 7-4 常见的逻辑运算符

运算符	表达式
&&（and，逻辑与）	${true && false}或${true and false}
||（or，逻辑或）	${true \|\| false}或${true or false}
!（not，逻辑非）	${!true}或${not true}

（4）empty 运算符

empty 运算符用于判断对象或变量是否为 null 或为空，格式如下。

```
${empty student}
```

其中，student 为要判断的变量或对象。

> 【脚下留心】
> 　　对象或变量为空与为 null 所表达的含义是不一样的，但用 empty 运算符判断的结果都为 True。

（5）条件运算符

EL 的条件运算符与 Java 的三元运算符几乎一样，格式如下。

```
${(1 < 2) ? a:b}
```

上述表达式的值为 a。因为表达式在判断"1<2"的时候，判断结果为 True，所以就执行了"a"，并且最终返回的也是"a"。如果将前面的表达式改为"1>2"，则结果为"b"。

8. EL 运算符的优先级

这里需要说明一点，EL 中的"()"与 Java 中的一样，都用于改变运算符的优先级。比如表达式：${1*2+3}，这里本应该先计算 1*2，再将得到的结果做加 3 的运算，但是如果给 2+3 加上括号，改为${1*（2+3）}，则会先计算 2+3 的和，再将和与 1 相乘。

常见的运算符按优先级从最高到最低的顺序排列如下。

① "[]"和"."。
② "()"。
③ "!""not"逻辑运算符与 empty 运算符。
④ "*""/""div""%""mod"。
⑤ "+""-"。
⑥ "<"">""<="">=""lt""gt""le""ge"比较运算符。
⑦ "==""!=""eq""ne"比较运算符。
⑧ "&&"或"and"逻辑运算符。
⑨ "||"或"or"逻辑运算符。
⑩ "?:"条件运算符。

9. 使用 EL 表达式从作用域中获取数据

在 JSP2.0 之前，只能使用如下代码访问系统作用域的值。

```
<%= session.getAttribute("name") %>
```

在使用 EL 表达式以后，就可以用如下代码访问同样的信息。

```
${name}
```

下面列出几种从不同作用域中获取数据的方法。
① 使用 EL 表达式从 request 作用域中获取数据。
JSP 脚本：<%=request.getAttribute（"name"）%>。
使用 EL 表达式替换上面的 JSP 脚本：${requestScope.name}。
② 使用 EL 表达式从 session 作用域中获取数据。
JSP 脚本：<%=session.getAttribute（"name"）%>。

使用 EL 表达式替换上面的 JSP 脚本：${sessionScope.name}。
③使用 EL 表达式从 application 作用域中获取数据。
JSP 脚本：<%=application.getAttribute("name")%>。
使用 EL 表达式替换上面的 JSP 脚本：${applicationScope.name}。

【多学一招】
　　EL 表达式可以和 JS 语句同时使用，也可以执行算术运算、逻辑运算、关系运算、条件运算。

10. EL 的隐式对象

EL 的隐式对象比 JSP 多，共有 11 个，所以使用 EL 获取数据比使用 JSP 要方便很多。

常见的隐式对象有 session、Cookie、header、headerValues、param 和 paramValues 等。如果要获取某个隐式对象内部的某个值，使用"${隐式对象名称["元素"]}"这种格式就可以。如要获取 HTTP 头部中"host"的值，就可这样写：

```
${header["host"]}
```

注意：不要把 JSP 的隐式对象和 EL 的隐式对象混淆，其中，pageContext 对象是 EL 和 JSP 所共有的。常见的 EL 隐式对象及其作用如表 7-5 所示。

表 7-5　常见的 EL 隐式对象及其作用

EL 隐式对象	作　　用
pageContext	与 JSP 的 pageContext 隐式对象功能相同，可以访问 JSP 的隐式对象
pageScope	访问 pageScope 范围的变量
requestScope	访问 requestScope 范围的变量
sessionScope	访问 sessionScope 范围的变量
applicationScope	访问 applicationScope 范围的变量
Param	获取 request 对象参数的单个值
paramValues	获取 request 对象参数的一个数值数组
header	将请求头名称映射到单个 Cookie 对象
headerValues	将请求头名称映射到一个数值数组
Cookie	将 Cookie 名称映射到单个 Cookie 对象
initParam	将上下文初始化参数名称映射到单个值

这些 EL 的隐式对象可分为三种类型。

（1）访问作用域范围的隐式对象

在 EL 中访问作用域范围的隐式对象有 4 个：applicationScope、sessionScope、requestScope、pageScope。这 4 个隐式对象分别可以获取不同域范围的信息，而不会获取其他域的信息。

【例 7-3】应用 EL 获取并输出 4 个隐式对象的属性值，参考代码如下。

第一步：新建 myServlet2。

```
package el.com;
```

```java
import java.io.IOException;
import javax.servlet.ServletContext;
import javax.servlet.ServletException;
import javax.servlet.annotation.WebServlet;
import javax.servlet.http.HttpServlet;
import javax.servlet.http.HttpServletRequest;
import javax.servlet.http.HttpServletResponse;
import javax.servlet.http.HttpSession;
@WebServlet("/myServlet2")
public class myServlet2 extends HttpServlet {
    private static final long serialVersionUID = 1L;
    protected void doGet(HttpServletRequest request,HttpServletResponse response) throws ServletException,IOException {
        response.setContentType("text/html;charset=UTF-8");
        request.setAttribute("userName","薛涛 request");
        HttpSession httpSession=request.getSession();
        httpSession.setAttribute("username","薛涛 session");
        ServletContext servletContext=request.getServletContext();
        servletContext.setAttribute("userName","薛涛 application");
        request.getRequestDispatcher("/scope.jsp").include(request,response);
    }
}
```

第二步：新建 scope.jsp。

```jsp
<%@ page language="java" contentType="text/html;charset=UTF-8"
    pageEncoding="UTF-8"%>
<!DOCTYPE html>
<html>
<head>
<meta charset="UTF-8">
<title>Insert title here</title>
</head>
<body>
<h3>应用 EL 获取并输出 4 个隐式对象的属性值</h3>
${ userName } <br/>
${ requestScope.userName } <br/>
${ sessionScope.userName } <br/>
${ applicationScope.userName } <br/>
<% pageContext.setAttribute("userName","薛涛 pageContext");%>
${ userName } <br/>
${ pageScope.userName } <br/>
</body>
</html>
```

得到 4 个隐式对象的列属性值如图 7-5 所示。

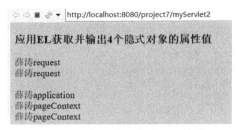

图 7-5　4 个隐式对象的列属性值

【多学一招】

若不指定作用范围，EL 将按照 pageContext、request、session、ServletContext 隐式对象的顺序，依次查找被绑定的属性。

（2）访问环境信息的隐式对象

在 EL 中访问环境信息的隐式对象有 6 个：Cookie、initParam、header、param、headerValues、paramValues。其中用得较多的是 param 和 paramValues，这两个隐式对象专门用于获取客户端访问 JSP 时传递的请求参数。

Cookie 隐式对象提供了对 Cookie 的访问。比如在 Cookie 中有一个名为 username 的值，要获取这个值可以通过如下方式。

```
${cookie.username}
```

initParam 隐式对象用于获取 web.xml 中一些初始参数的值。比如，web.xml 中的参数如下。

```
<context-param>
    <param-username>admin</param-username>
    <param-password>123456</param-password>
</context-param>
```

则在 JSP 页面中，可以直接应用 EL 获取 admin 所对应的密码值，代码如下。

```
${initParam.admin}
```

得到结果为 123456。

【例 7-4】应用 EL 获取并输出请求参数的值的代码如下。

第一步：创建一个 MyServlet3。

教学视频

```
package el.com;
import java.io.IOException;
import javax.servlet.ServletException;
import javax.servlet.annotation.WebServlet;
import javax.servlet.http.HttpServlet;
import javax.servlet.http.HttpServletRequest;
import javax.servlet.http.HttpServletResponse;
@WebServlet("/MyServlet3")
public class MyServlet3 extends HttpServlet {
    private static final long serialVersionUID = 1L;
    protected void doGet(HttpServletRequest request,HttpServletResponse response)
throws ServletException,IOException {
```

```
            response.setContentType("text/html;charset=UTF-8");
            String name="鱼玄机";
            String paraPath="parm.jsp?userName="+name+"&hobby=sport&hobby=music";
            request.getRequestDispatcher(paraPath).include(request,response);
            return;
        }
}
```

第二步：创建一个 parm.jsp 页面。

```jsp
<%@ page language="java" contentType="text/html;charset=UTF-8"
    pageEncoding="UTF-8"%>
<!DOCTYPE html>
<html>
<head>
<meta charset="UTF-8">
<title>Insert title here</title>
</head>
<body bgcolor="pink">
<h3>应用 EL 获取并输出请求参数的值</h3>
<%request.setCharacterEncoding("UTF-8");%>
${ param } <br/>
${ paramValues } <br/>
${ param.userName } <br/>
${ paramValues.hobby } <br/>
${ paramValues.hobby[0] } <br/>
${ paramValues.hobby[1] } <br/>
</body>
</html>
```

运行结果如图 7-6 所示。

图 7-6　获取并输出请求参数的值

（3）pageContext 隐式对象

应用 EL 可获取 pageContext 隐式对象，下面通过例 7-5 说明。

【例 7-5】应用 EL 获取 pageContext 隐式对象并输出请求参数的值，参考代码如下。

第一步：创建一个 MyServlet4。

教学视频

```java
package el.com;
import java.io.IOException;
```

```java
import javax.servlet.ServletException;
import javax.servlet.annotation.WebServlet;
import javax.servlet.http.HttpServlet;
import javax.servlet.http.HttpServletRequest;
import javax.servlet.http.HttpServletResponse;
@WebServlet("/MyServlet4")
public class MyServlet4 extends HttpServlet {
    private static final long serialVersionUID = 1L;
    protected void doGet(HttpServletRequest request,HttpServletResponse response) throws ServletException,IOException {
        response.setContentType("text/html;charset=UTF-8");
        request.getRequestDispatcher("pageContext.jsp").include(request,response);
        return;
    }
}
```

第二步：创建一个 pageContext.jsp 页面。

```jsp
<%@ page language="java" contentType="text/html;charset=UTF-8"
    pageEncoding="UTF-8"%>
<!DOCTYPE html>
<html>
<head>
<meta charset="UTF-8">
<title>Insert title here</title>
</head>
<body>
<%request.setCharacterEncoding("UTF-8");%>
<h3>应用 EL 获取 pageContext 对象</h3>
${ pageContext['request'] } <br/>
${ pageContext.session } <br/>
${ pageContext['servletContext'] } <br/>
${ pageContext.request.servletContext } <br/>
${ pageContext.response } <br/>
${ pageContext.request.contextPath } <br/>
</body>
</html>
```

运行结果如图 7-7 所示。

图 7-7 输出请求参数的值

任务实施

第一步：创建一个 forminfo.jsp 页面。

```jsp
<%@ page language="java" contentType="text/html;charset=UTF-8"
    pageEncoding="UTF-8"%>
<!DOCTYPE html>
<html>
<head>
<meta charset="UTF-8">
<title>Insert title here</title>
</head>
<body bgcolor="pink">
<b>
请填写用户信息<br/>
<form action="el.jsp" method="get">
用户名:<input type="text" name="username"/> <br/>
密码:<input type="password" name=pwd/> <br/>
性别:<input type="radio" value="男" name="sex"/>男
<input type="radio" value="女" name="sex"/>女<br/>
爱好:<input type="checkbox" value="体育" name="hobby"/>体育
<input type="checkbox" value="美术" name="hobby"/>美术
<input type="checkbox" value="音乐" name="hobby"/>音乐
<input type="checkbox" value="旅游" name="hobby"/>旅游<br/>
<input type="submit" value="提交">
<input type="reset" value="重置">
</b> <br/>
<textArea cols="30" rows="3">
该表单信息将提交给el.jsp页面,通过EL表达式获取表单的信息
</textArea>
</form>
</body>
</html>
```

第二步：创建一个 JavaBean，类名为 UserInfo.java。

```java
package el.bean;
public class UserInfo {
private String username;
private String pwd;
private String sex;
private String[] hobby=null;
public String getUsername() {
    return username;
}
public void setUsername(String username) {
    this.username = username;
}
```

```java
    public String getPwd() {
        return pwd;
    }
    public void setPwd(String pwd) {
        this.pwd = pwd;
    }
    public String getSex() {
        return sex;
    }
    public void setSex(String sex) {
        this.sex = sex;
    }
    public String[] getHobby() {
        System.out.println(hobby[0]);
        return hobby;
    }
    public void setHobby(String[] hobby) {
        this.hobby = hobby;
    }
}
```

第三步：创建一个 el.jsp 页面，用于获取表单的信息。

```jsp
<%@ page language="java" contentType="text/html;charset=UTF-8"
    pageEncoding="UTF-8"%>
<%@ page import="el.bean.UserInfo" %>
<!DOCTYPE html>
<html>
<head>
<meta charset="UTF-8">
<title>Insert title here</title>
</head>
<body bgcolor="pink">
<%
request.setCharacterEncoding("UTF-8");
UserInfo user=new UserInfo();
user.setUsername((String)request.getParameter("username"));
user.setPwd((String)request.getParameter("pwd"));
user.setSex(request.getParameter("sex"));
user.setHobby((String[])request.getParameterValues("hobby"));
session.setAttribute("userform",user);
%>
<b>
用户显示:<br>
用户名:${sessionScope.userform.username}<br>
密码:${sessionScope.userform.pwd}<br>
性别:${sessionScope.userform.sex}<br>
爱好:${sessionScope.userform.hobby[0]}  ${sessionScope.userform.hobby
```

```
[1]}  ${sessionScope.userform.hobby[2]}<br>
    </b>
    </body>
    </html>
```

任务拓展

1. 应用 EL 获取 Cookie 对象的信息

在 JSP 开发中经常需要获取客户端的 Cookie 对象的信息，为此，EL 提供了 Cookie 隐式对象。该对象是一个集合了所有 Cookie 信息的 Map 集合，Map 集合中元素的键为 Cookie 的名称，值为对应的 Cookie 对象。

${cookie.名称}：根据名称获取对应的 Cookie 对象，如${cookie.JSESSIONID}。

${cookie.名称.value}：根据名称获取对应 Cookie 对象的值，如${cookie.JSESSIONID.value}。

【例 7-6】应用 EL 获取 Cookie 对象的信息，参考代码如下。

```
<%@ page language="java" contentType="text/html;charset=UTF-8"
    pageEncoding="UTF-8"%>
<!DOCTYPE html>
<html>
<head>
<meta charset="UTF-8">
<title>Insert title here</title>
</head>
<body>
<%
Cookie cookie=new Cookie("userName","Username in cookie");
response.addCookie(cookie);
%>
Cookie 对象的信息:<br/>
${cookie.userName} <br/>
Cookie 对象的名称和值:<br/>
${cookie.userName.name}= ${cookie.userName.value}<br/>
</body>
</html>
```

运行结果如图 7-8 所示。

图 7-8 获取 Cookie 对象的信息

2. 应用 EL 获取 initParam 对象的信息

initParam 对象可用于获取 Web 应用初始化参数的值，下面通过例 7-7 具体讲解 initParam

对象的使用。

【例 7-7】应用 EL 获取 initParam 对象的信息，代码如下。

第一步：在 web.xml 文件中设置一个初始化参数 author。

教学视频

```xml
<context-param>
    <param-name>author</param-name>
    <param-value>廖丽</param-value>
</context-param>
```

第二步：创建一个 ElInitParam.jsp 页面。

```jsp
<%@ page language="java" contentType="text/html;charset=UTF-8"
    pageEncoding="UTF-8"%>
<!DOCTYPE html>
<html>
<head>
<meta charset="UTF-8">
<title>Insert title here</title>
</head>
<body bgcolor="pink">
author 的值为:${initParam.author}
</body>
</html>
```

运行结果如图 7-9 所示。

图 7-9 获取 initParam 对象的信息

任务 2　应用 JSTL 实现商品展示

 任务演示

创建 Map 集合，实现商品列表展示，展示效果如图 7-10 所示。

 知识准备

1. JSTL 的概念

JSTL 全称是 Java server pages Standard Tag Library，译作 JSP 标准标签库，是由 JCP（Java Community Process）制定的标准规范，它主要为 Java Web 开发人员提供一个标准、通用的标签库，并由 Apache 的 Jakarta 小组维护。开发人员可以利用标签取代 JSP 页面上的 Java 代码，从而提高程序的可读性，降低程序的维护难度。

JSTL 标签是基于 JSP 页面设计的，这些标签可以插入 JSP 代码中。JSTL 本质上也是提

图 7-10　商品列表展示效果

前定义好的一组标签，这些标签封装了不同的功能，在页面上调用标签，就等于调用了封装起来的功能。使用 JSTL 是为了弥补 HTML 表的不足，规范自定义标签的使用。在告别使用 model1 模式开发应用程序后，人们开始注重软件的分层设计，他们不希望在 JSP 页面中出现 Java 逻辑代码，同时也考虑到自定义标签的开发难度较大和不利于技术标准化，于是产生了自定义标签库。在此仅简要介绍 JSTL 的 5 个标签库。

①核心标签库：最常用的标签库，为日常工作提供通用支持，如判断、循环等。
②国际化（I18N）与格式化标签库：支持多种语言。
③SQL 标签库：可用于操作数据库。
④XML 标签库：可用于处理 XML 文件，包括 XML 节点解析、迭代等。
⑤函数标签库：通过在 EL 表达式中调用函数标签库中的函数来实现特定操作。
以上 JSTL 对应的 taglib 指令如表 7-6 所示。

表 7-6　常见标签库对应的 taglib 指令

标签库	URL	前缀	举例
核心标签库	http://java.sun.com/jstl/core	c	<c:tagname...>
国际化（I18N）与格式化标签库	http://java.sun.com/jstl/fmt	fmt	<fmt:tagname...>
SQL 标签库	http://java.sun.com/jstl/sql	sql	<sql:tagname...>
XML 标签库	http://java.sun.com/jstl/x	x	<x:tagname...>
函数标签库	http://java.sun.com/jstl/fn	fn	<fn:tagname...>

【温馨提示】
　　不必将 5 个标签库都记住，只需要学会核心标签库的使用就可以了，其他标签库可在需要时再查阅相关资料。

2. 下载和安装 JSTL

（1）下载 JSTL 包

第一步：进入 Apache 官网首页，如图 7-11 所示，单击"Standard"按钮，进入 JSTL 版本界面。

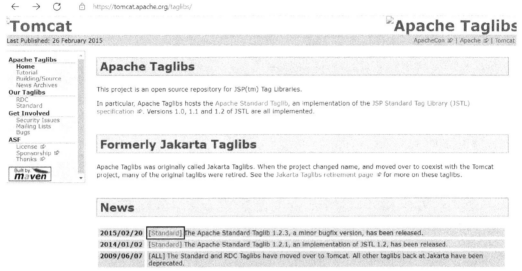

图 7-11　Apache 官网首页

第二步：进入 JSTL 版本界面后，单击对应版本的"download"按钮进入下载界面，如图 7-12 所示。

图 7-12　JSTL 版本界面

第三步：进入下载界面后，会出现 JSTL 各个 jar 包的下载链接，单击需要下载的 jar 包即可，下载界面如图 7-13 所示。

> 【多学一招】
> Impl：taglibs-standard-impl-1.2.5.jar，JSTL 实现类库。
> Spec：taglibs-standard-spec-1.2.5.jar，JSTL 标准接口。
> EL：taglibs-standard-jstlel-1.2.5.jar，JSTL1.0 标签，与 EL 相关。
> Compat：taglibs-standard-compat-1.2.5.jar，兼容版本。

图 7-13　下载界面

（2）导入 JSTL 包

如果要在 JSP 页面中使用 JSTL，就必须将 JSTL 标签库添加到 Web 应用中，将下载好的 taglibs-standard-impl-1.2.5.jar 和 taglibs-standard-spec-1.2.5.jar 粘贴到 project7 项目的 WebContent 目录下的 WEB-INF 文件夹下的 lib 文件夹里，如图 7-14 所示。下面还需要在 Eclipse IDE 中配置和引入的两个包才可以正常使用，如图 7-15 所示，最后单击"Apply"按钮实现包的引入。

图 7-14　粘贴 jar 包的路径

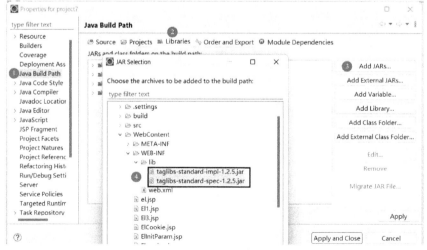

图 7-15　配置和引入包

在包引入完成后，还需测试是否安装成功。在此新建一个 test.jsp 文件，再使用 JSTL 的 JSP 导入标签库，代码如下。

```
<%@taglib uri="http://java.sun.com/jsp/jstl/core" prefix="c"%>
```

【脚下留心】
　　uri：标签库的唯一标识，为了保证唯一性，其值通常是一个 URL 或者一个电子邮箱地址。
　　prefix：标签库的前缀，用于区分不同标签库中的同名标签，这里将值设置为 c，那么后续使用的就是<c: xxx>。

【例 7-8】测试 JSTL 是否安装成功，代码如下。

教学视频

```
<%@ page language="java" contentType="text/html;charset=UTF-8"
    pageEncoding="UTF-8"%>
<%@ taglib prefix="c" uri="http://java.sun.com/jsp/jstl/core" %>
<!DOCTYPE html>
<html>
<head>
<meta charset="UTF-8">
<title>Insert title here</title>
</head>
<body bgcolor="pink">
<c:out value="JSTL 能正常使用"></c:out>
</body>
</html>
```

JSTL 安装成功的运行结果如图 7-16 所示。

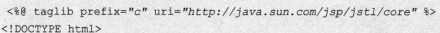

图 7-16　JSTL 安装成功的运行结果

3. JSTL 的核心标签库

JSTL 包含 5 个标签库，其中的核心标签库包含了 Web 应用中的通用操作标签，下面对核心标签库中的常用标签进行详细讲解。

（1）表达式标签

表达式标签包括<c:out>、<c:set>、<c:remove>标签，现在分别介绍它们的功能和语法。

① <c:out>标签

功能：用来显示数据对象（如字符串、表达式）的内容和结果。

格式： `<c:out value="this is JSTL" escapeXml="true|false" default="defaultValue"/>`。

参数说明：

value：指定将要输出的表达式。

default：表示当 value 的值为 null 时，将输出默认值。

escapeXml：确定是否应将结果中的字符（如<、>、&、'、" 等特殊的符号）转换为字符实体代码，默认值为 True，即要转换为实体代码。如：字符"<"可以转换为"<"。

【例 7-9】<c:out>标签的应用如下。

```jsp
<%@ page language="java" contentType="text/html;charset=UTF-8"
    pageEncoding="UTF-8"%>
 <%@ taglib prefix="c" uri="http://java.sun.com/jsp/jstl/core" %>
<!DOCTYPE html>
<html>
<head>
<meta charset="UTF-8">
<title>out 标签的使用</title>
</head>
<body bgcolor="pink">
<!-- value 属性赋值为字符串 -->
<li>(1)<c:out value="Java 动态网站学习"></c:out></li>
<!-- 字符串中有转义字符,但在默认情况下没有转换 -->
<li>(2)<c:out value="&lt 未使用字符转义&gt" /></li>
<!-- 改变 escapeXml 属性后输出的转义字符 -->
<li>(3)<c:out value="&lt 使用字符转义&gt" escapeXml="false"></c:out></li>
<!-- 设定了默认值,从 EL 表达式${null}得到空值,所以直接输出设定的默认值 -->
<li>(4)<c:out value="${null}">使用了默认值</c:out></li>
<!-- 未设定默认值,输出结果为空 -->
<li>(5)<c:out value="${null}"></c:out></li>
</body>
</html>
```

教学视频

<c:out>标签的应用结果如图 7-17 所示。

图 7-17 <c:out>标签的应用结果

② <c:set>标签

功能：主要用于将变量存取于 JSP 范围中或 JavaBean 属性中。

格式：`<c:set var="varName" value = "value" scope =" page|request|session|application"/>`。

参数说明：

var：指定创建的变量的名称，以存储标签中指定的 value 值。

value：指定表达式。

scope：指定变量的生命周期，默认值为 page。

【例 7-10】使用<c:set>存取值，代码如下。

```jsp
<%@ page language="java" contentType="text/html;charset=UTF-8"
    pageEncoding="UTF-8"%>
 <%@ taglib prefix="c" uri="http://java.sun.com/jsp/jstl/core" %>
<!DOCTYPE html>
```

教学视频

```
<html>
<head>
<meta charset="UTF-8">
<title>set 标签的使用</title>
</head>
<body bgcolor="pink">
<li>把一个值放入 session 中。<c:set value="coo" var="name1" scope="session"></c:set>
<li>从 session 中得到值:${sessionScope.name1 }
<li>把另一个值放入 application 中。<c:set var="name2" scope="application"> olive
</c:set>
<li> 使用 out 标签和 EL 表达式嵌套得到值:
<c:out value="${applicationScope.name2}">未得到 name 的值</c:out></li>
<li>未指定 scope 的范围,会从不同的范围内查找得到相应的值:${name1 }、${name2 }
</body>
</html>
```

使用<c:set>存取值的运行结果如图 7-18 所示。

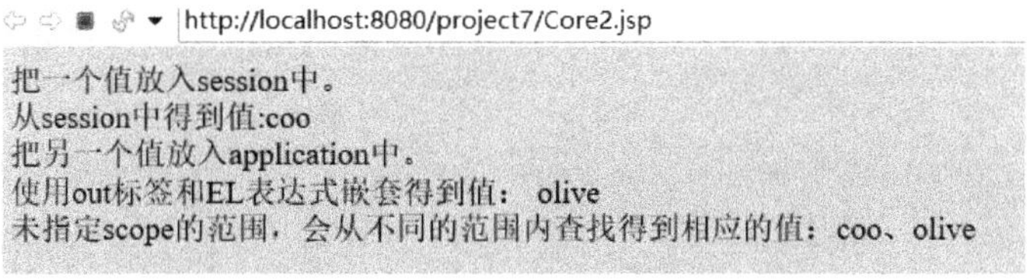

把一个值放入session中。
从session中得到值:coo
把另一个值放入application中。
使用out标签和EL表达式嵌套得到值： olive
未指定scope的范围，会从不同的范围内查找得到相应的值： coo、olive

图 7-18　使用<c:set>存取值的运行结果

③ <c:remove>标签

作用：主要用来从指定的 JSP 范围内移除指定的变量。

格式：`<c:remove var = "varName" [ scope = "page|request|session|application"]/>`。

参数说明：

scope：需要移除的变量的所在范围。

var：需要移除的变量或者对象属性的名称。如果没有 scope 属性，即采用默认值，相当于调用 pageContext. removeAttribute(varName)方法；如果指定了变量所在范围，那么系统会调用 pageContext. removeAttribute(varName,scope)方法。

【例 7-11】<c:remove>标签的应用，代码如下。

```
<%@ page language="java" contentType="text/html;charset=UTF-8"
    pageEncoding="UTF-8"%>
<%@ taglib prefix="c" uri="http://java.sun.com/jsp/jstl/core" %>
<!DOCTYPE html>
<html>
<head>
<meta charset="UTF-8">
<title>catch 标签</title>
</head>
<body bgcolor="pink">
<!-- 使用 set 标签向 session 中插入三个值:name 值为张三、age 值为 20、sex 值为男 -->
<li><c:set var="name" scope="session">张三</c:set>
```

```
            <li><c:set var="age" scope="session">20</c:set>
            <li><c:set var="sex" scope="session">男</c:set>
            <!-- 使用 out 标签和 EL 表达式输出 name、age、sex 的值 -->
            <li><c:out value="${sessionScope.name}"></c:out>
            <li><c:out value="${sessionScope.age}"></c:out>
            <li><c:out value="${sessionScope.sex}"></c:out>
            <!-- 使用 remove 标签移除 age 的值 -->
            <li><c:remove var="age"/>
            <!-- 使用 out 标签和 EL 表达式输出 name、age、sex 的值 -->
            <li><c:out value="${sessionScope.name}"></c:out>
            <li><c:out value="${sessionScope.age}"></c:out>
            <li><c:out value="${sessionScope.sex}"></c:out>
</body>
</html>
```

<c:remove>标签的应用结果如图 7-19 所示。

图 7-19 <c:remove>标签的应用结果

（2）流程控制标签

流程控制标签包括<c:if>、<c:choose>、<c:when>、<c:otherwise>等。流程控制标签根据其 test 属性值决定是否执行其标签体中的内容。

① <c:if>标签

作用：同程序中的 if 语句的作用相同，用来实现条件控制。

格式如下：

```
<c:if test="条件" var="name" [scope="page|request|session|application"]>
```

参数说明：

test：用于指定条件。

var：用于保存根据 test 指定的条件表达式返回的值（即 True 或 False）。

scope：指定 var 的范围。

【例 7-12】<c:if>标签的应用，代码如下。

教学视频

```
<%@ page language="java" contentType="text/html;charset=UTF-8"
    pageEncoding="UTF-8"%>
<%@ taglib prefix="c" uri="http://java.sun.com/jsp/jstl/core" %>
<!DOCTYPE html>
<html>
```

```
<head>
<meta charset="UTF-8">
<title>if 标签应用</title>
</head>
<body bgcolor="pink">
<h4>if 标签示例</h4>
 <hr>
<c:set value="1" var="count" property="count"/>
<c:if test="${count==1}">
    四大发明是中国古代创新的智慧成果和科学技术,包括造纸术、指南针、火药、印刷术
</c:if>
</body>
</html>
```

<c:if>标签的应用结果如图 7-20 所示。

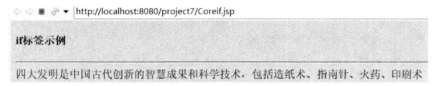

图 7-20 <c:if>标签的应用结果

② <c:choose>标签、<c:when>标签、<c:otherwise>标签

通常情况下，<c:choose>标签、<c:when>标签、<c:otherwise>标签是一起使用的，<c:choose>标签被当作<c:when>标签和<c:otherwise>标签的父标签来使用。在<c:choose>标签中嵌入了多个<c:when>子标签，每个<c:when>子标签有一个 test 属性，如果 test 的属性值为 True，则执行<c:when>标签体，其中的<c:when>子标签和<c:otherwise>子标签相当于 Java 中 switch 语句的 case 和 default。其格式如下。

```
<c:choose>
    <c:when test = "判断条件1">
        内容1
    </c:when>
    <c:when test = "判断条件2">
        内容2
    </c:when>
    …
    <c:otherwise>
        内容N
    </c:otherwise>
</c:choose>
```

参数说明：

<c:choose>标签：只能和<c:when>标签共同使用。

<c:when>标签：用于条件判断，一般情况下和<c:choose>标签共同使用。

<c:otherwise>标签：不含参数，只能跟<c:when>标签共同使用，并且只允许在嵌套中出现一次。

【例 7-13】通过键盘输入一个数字,如果这个数字大于 0 且小于 6,则打印"正在营业";如果这个数字等于 6,则输出"歇业一天";如果这个数字等于 7,则输出"照常营业";如果这个数字是其他情况,则输出"输入有误"。参考代码如下。

教学视频

```
<%@ page language="java" contentType="text/html;charset=UTF-8"
    pageEncoding="UTF-8"%>
<%@ taglib prefix="c" uri="http://java.sun.com/jsp/jstl/core" %>
<!DOCTYPE html>
<html>
<head>
<meta charset="UTF-8">
<title>Insert title here</title>
</head>
<body bgcolor="pink">
<h4>choose 及其嵌套标签示例</h4>
    <hr>
            <c:set var="n">4</c:set>
            <c:choose>
            <c:when test="${n>0&&n<6}">
                        正在营业
            </c:when>
            <c:when test="${n==6}">
                        歇业一天
            </c:when>
            <c:when test="${n==7}">
                        照常营业
            </c:when>
            <c:otherwise>
                        输入有误
            </c:otherwise>
            </c:choose>
</body>
</html>
```

程序运行结果如图 7-21 所示。

图 7-21　程序运行结果

(3) 循环标签

循环标签包括<c:forEach>标签和<c:forTokens>标签。

① <c:forEach>标签

作用:该标签可根据循环条件遍历集合中的元素。

格式：

```
<c:forEach items = "collection" var = "varNane" [varstatus = "varStatusName" ]
[begin="begin"] [end="end"][step="step"] >
内容体
</c:forEach>
```

参数说明：

var：指保存在集合类（collection）中的对象名称。

items：指要循环的集合类名。

varstatus：用于存储循环的状态信息，可以访问到循环自身的信息。

begin：如果指定了 begin 值，就表示从 items[begin]开始迭代；如果没有指定 begin 值，则从集合的第一个值开始迭代。

end：表示循环到集合的第 end 位时结束，如果没有指定 end 值，则表示循环到集合的最后一位。

step：指定循环的步长。

【例 7-14】应用<c:forEach>标签输出唐代四大书法家，代码如下。

```jsp
<%@ page language="java" contentType="text/html;charset=UTF-8"
    pageEncoding="UTF-8"%>
<%@page import="java.util.List"%>
<%@page import="java.util.ArrayList"%>
<%@ taglib prefix="c" uri="http://java.sun.com/jsp/jstl/core" %>
<!DOCTYPE html>
<html>
<head>
<meta charset="UTF-8">
<title>forEach 标签</title>
</head>
<body bgcolor="pink">
<h4><c:out value="forEach 实例"/></h4>
    <hr>
<!-- 创建了一个集合对象 a,并添加元素 -->
 <%
 List a=new ArrayList();
  a.add("褚遂良");
  a.add("欧阳询");
  a.add("颜真卿");
  a.add("柳公权");
 //使用 setAttribute()方法把集合存入 request 范围内
  request.setAttribute("a",a);
 %>
<B><c:out value="从集合开始遍历,到集合结束为止" /></B><br>
 <c:forEach var="fuwa" items="${a}">
  <c:out value="${fuwa}"/><br>
</c:forEach>
<!-- 指定从集合的第二个(index 值为 1)元素开始,到第四个(index 值为 3)元素截止,并指定 step 为 2
```

```
即每隔两个元素遍历一次 -->
    <B><c:out value="指定begin和end的迭代:" /></B><br>
    <c:forEach var="fuwa" items="${a}" begin="1" end="3" step="2">
     <c:out value="${fuwa}" /><br>
    </c:forEach>
    <B><c:out value="输出整个迭代的信息:" /></B><br>
    <c:forEach var="fuwa" items="${a}" begin="2" end="3" step="1" varStatus="s">
     <c:out value="${fuwa}" />的四种属性:<br>
      所在位置,即索引:<c:out value="${s.index}" /><br>
      总迭代次数:<c:out value="${s.count}" /><br>
      是否为第一个位置:<c:out value="${s.first}" /><br>
      是否为最后一个位置:<c:out value="${s.last}" /><br>
    </c:forEach>
</body>
</html>
```

程序运行结果如图 7-22 所示。

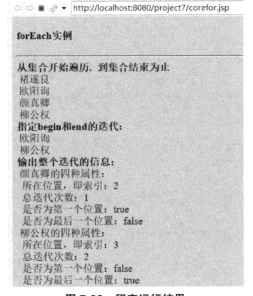

图 7-22　程序运行结果

从程序运行结果可以看到,如果不指定 begin 和 end 的值,则循环从集合的第一个元素开始,遍历到集合的最后一个元素结束。

② <c:forTokens>标签

作用:该标签用于浏览字符串,并根据指定的字符截取字符串。与<c:forEach>标签一样,<c:forTokens>标签也可以指定 begin、end、step 的属性值。

格式:

```
<c:forTokens items="stringOfTokens" delims="delimiters" var="varName" [varStatus=
"varStatusName"] [begin= "begin"] [end = "end"] [step = "step"]>
    内容体
</c:forTokens>
```

参数说明：

var：表示循环的参数名称。

items：指定进行标签化的字符串。

varStatus：表示每次循环的状态信息。

delims：指定分隔符来分隔 items 指定的字符串。

begin：表示循环开始的位置。

end：表示循环结束的位置。

step：表示循环的步长。

 任务实施

第一步：创建 User 的 bean 类。

```java
package el.bean;
public class User {
int no;
String name;
double score;
public User(int no,String name,double score) {
    super();
    this.no = no;
    this.name = name;
    this.score = score;
}
public int getNo() {
    return no;
}
public void setNo(int no) {
    this.no = no;
}
public String getName() {
    return name;
}
public void setName(String name) {
    this.name = name;
}
public double getScore() {
    return score;
}
public void setScore(float score) {
    this.score = score;
}
}
```

第二步：创建 ListServlet。

```java
package el.com;
```

```java
import el.bean.Phone;
import javax.servlet.ServletException;
import javax.servlet.annotation.WebServlet;
import javax.servlet.http.HttpServlet;
import javax.servlet.http.HttpServletRequest;
import javax.servlet.http.HttpServletResponse;
import java.io.IOException;
import java.util.ArrayList;
import java.util.List;
@WebServlet("/ListServlet")
public class ListServlet extends HttpServlet{
    protected void doPost(HttpServletRequest request,HttpServletResponse response) throws ServletException,IOException {
    }
    protected void doGet(HttpServletRequest request,HttpServletResponse response) throws ServletException,IOException
    {
        Phone phone=new Phone();
        phone.setName("iphone6");
        phone.setId(001);
        phone.setImage("image/a1.jpg");
        phone.setPrice("3900");
        Phone phone1=new Phone();
        phone1.setName("坚果pro");
        phone1.setId(002);
        phone1.setPrice("1799");
        phone1.setImage("image/a2.jpg");
        Phone phone2=new Phone();
        phone2.setName("vivo x9");
        phone2.setPrice("2345");
        phone2.setId(003);
        phone2.setImage("image/a3.jpg");
        Phone phone3=new Phone();
        phone3.setName("oppo A57");
        phone3.setId(004);
        phone3.setPrice("1399");
        phone3.setImage("image/a4.jpg");
        Phone phone4=new Phone();
        phone4.setName("诺基亚6");
        phone4.setId(005);
        phone4.setPrice("1699");
        phone4.setImage("image/a5.jpg");
        Phone phone5=new Phone();
        phone5.setName("小米MIX");
        phone5.setId(006);
        phone5.setPrice("3999");
        phone5.setImage("image/a6.jpg");
        List list=new ArrayList<>();
```

```
            list.add(phone);
            list.add(phone1);
            list.add(phone2);
            list.add(phone3);
            list.add(phone4);
            list.add(phone5);
            request.setAttribute("list",list);
            request.getRequestDispatcher("/phone_list.jsp").forward(request,
response);
        }
    }
```

第三步：创建 phone_list.jsp。

```
<%@ page language="java" contentType="text/html;charset=UTF-8"
    pageEncoding="UTF-8"%>
<%@taglib prefix="c" uri="http://java.sun.com/jsp/jstl/core" %>
<!DOCTYPE html>
<html>
<head>
<meta charset="UTF-8">
<title>商品列表</title>
</head>
<body>
<c:forEach items="${list}" var="phone">

    <div style="height:250px">
        <img src="${phone.image}" width="170" height="200" style="display:inline-block;">
        <p>
            <a href="product_info.html" style='color:green'>${phone.name}</a>
        </p>
        <p>
            <font color="#FF0000">商城价:&yen;${phone.price}</font>
        </p>
    </div>
</c:forEach>
</body>
</html>
```

任务拓展

通常需要在 JSP 页面中实现 URL 的重写及重定向等特殊功能。为了实现这些功能，核心标签库提供了相应标签，包括<c:param>标签、<c:redirect>标签和<c:url>标签。下面将详细介绍这 3 个标签的使用方法。

（1）<c:param>标签

作用：用于获取 URL 地址中的附加参数。

<c:param>标签有两种语法格式，分别如下。

① 使用 value 属性指定参数的值。

```
<c:param name="name" value="value"
```

② 在标签体中指定参数的值。

```
<c:param name="name">
    parameter value
</c:param>
```

参数说明：

name：用于指定参数的名称。

value：用于指定参数的值。当使用<c:param>标签为一个 URL 地址附加参数时，它会自动对参数值进行 URL 编码。例如，如果传递的参数值为"你好"，则需先将其转换为"%E4%BD% A0%E5%A5%BD"后再附加到 URL 地址后面。

（2）<c:redirect>标签

作用：用于重定向。

格式一：

```
<c:redirect url="url"[context="context"]>
```

格式二：

```
<c:redirect url="url"[context="context"]>
<c:param name="name1" value="value1">
</c:redirect>
```

参数说明：

url：指定重定向页面的地址，可以是一个 string 类型的绝对地址或相对地址。

（3）<c:url>标签

作用：用于按特定的规则重新构建 URL。

格式一：指定一个 URL，不做修改，可以选择把该 URL 存储在 JSP 的不同范围中。

```
<c:url value="value" [var="name"][scope="page|request|session|application"]
[context="context"]/>
```

格式二：给 URL 加上指定参数及参数值，可以选择用参数 name 存储该 URL。

```
<c:url value="value" [var="name"][scope="page|request|session|application"]
[context="context"]>
<c:param name="参数名" value="值">
</c:url>
```

【例 7-15】URL 标签的应用代码如下。

```
<%@ page language="java" contentType="text/html;charset=UTF-8"
    pageEncoding="UTF-8"%>
```

```
<%@ taglib prefix="c" uri="http://java.sun.com/jsp/jstl/core" %>
<!DOCTYPE html>
<html>
<head>
<meta charset="UTF-8">
<title>Insert title here</title>
</head>
<body>
<c:url var="myURL" value="http://localhost:8080/project7/c_out1.jsp">
<c:param name="username" value="李四" />
</c:url>
<a href="${myURL}">c_out1.jsp</a><br/>
使用相对路径构造 url:<br/>
<c:url var="myURL" value="c_out1.jsp?username=li"/>
<a href="${myURL}">c_out1.jsp</a>
</body>
</html>
```

URL 标签的应用结果如图 7-23 所示。

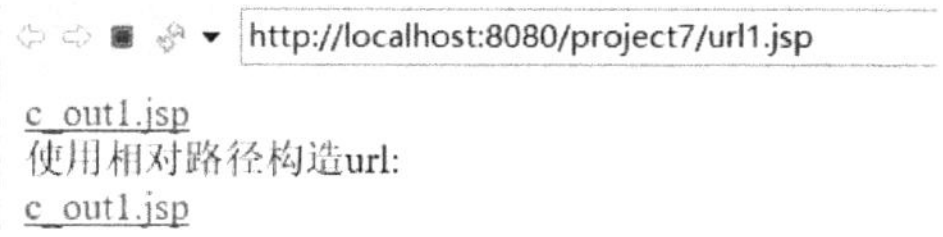

图 7-23　URL 标签的应用结果

【例 7-16】根据参数请求显示不同的页面。

通过键盘模拟爱情剧中的场景：从键盘输入一个数字，如果数字是 1314，则输出"在一起一辈子"；如果数字是 520，则输出"我们结婚吧"；如果不是上面两种情况，则什么都不输出。

```
<%@ page language="java" contentType="text/html;charset=UTF-8"
    pageEncoding="UTF-8"%>
<%@ taglib prefix="c" uri="http://java.sun.com/jsp/jstl/core" %>
<!DOCTYPE html>
<html>
<head>
<meta charset="UTF-8">
<title>Insert title here</title>
</head>
<body>
<c:url var="myURL" value="http://localhost:8080/project7/c_out1.jsp">
<c:param name="username" value="李四" />
</c:url>
<a href="${myURL}">c_out1.jsp</a><br/>
使用相对路径构造 url:<br/>
<c:url var="myURL" value="c_out1.jsp?username=li"/>
<a href="${myURL}">c_out1.jsp</a>
```

```
    </body>
</html>
```

当服务启动后，在浏览器地址栏中输入地址"http://localhost:8080/project7/url2.jsp?action=1314"访问 url2.jsp 页面。

程序运行结果如图 7-24 所示。

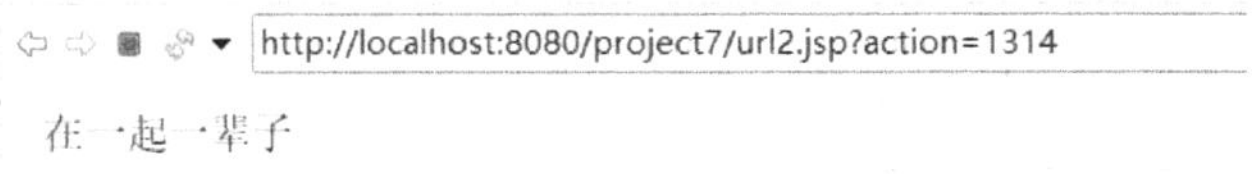

图 7-24　根据参数请求显示不同页面的程序运行结果

项 目 实 训

实训一　应用 EL 表达式显示"好好学习，天天向上！"

要求：应用 EL 表达式在页面上展示从后端 Servlet 传来的"好好学习，天天向上！"

实训二　应用 EL 表达式设置页面的背景色

要求：应用 EL 表达式的作用域与运算符来设置页面的背景色。

实训三　遍历集合中的元素

要求：应用 JSTL、HTML 展示数据，效果如图 7-25 所示。

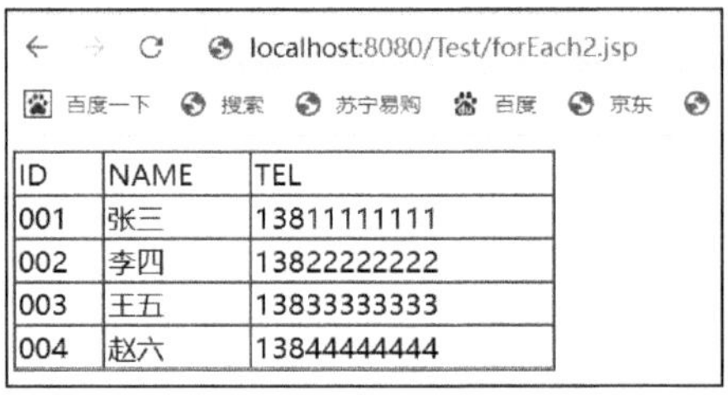

图 7-25　应用 JSTL、HTML 展示数据的效果

实训四　应用 EL 表达式得到一个计算器

要求：应用 EL 表达式得到一个计算器。效果如图 7-26 所示。

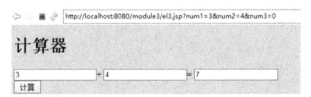

图 7-26　应用 EL 表达式得到一个计算器

课 后 练 习

一、填空题

1. EL 的_____隐式对象，代表 application 域中用于保存属性的 Map 对象。
2. _____是用于获取 Cookie 信息的隐式对象。
3. <c:forEach>标签可以迭代集合对象中的元素，包含 Set、List、Map 和_____等。
4. EL 的_____对象用于获取请求参数的某个值。
5. 如果要在 JSP 页面中导入核心标签库，需要使用_____指令。

二、选择题

1. 下列 JSTL 标签中，不属于流程控制标签的是（　　）。
A. <c:set>　　　　B. <c:choose>　　　　C. <c:when>　　　　D. <c:if>
2. 下列选项中，符合正确的 EL 表达式语法的是（　　）。
A. <username>　　　B. {username}　　　C. （username）　　　D. [username]
3. 下列选项中，不属于EL关键字的是（　　）
A. and　　　　　　B. or　　　　　　　C. not　　　　　　D. no
4. 下列关于<c:out>标签的说法中，错误的是（　　）。
A. <c:out>标签用于输出数据
B. <c:out>标签能够实现类似于 JSP 表达式的功能
C. <c:out>标签的 value 属性用于指定要输出的数据
D. <c:out>标签的 value 属性不能是 EL 表达式
5. 关于 EL 的隐式对象，下列说法中错误的是（　　）。
A. pageScope 可以取出最小的域对象 pageContext 中的参数
B. request、session 是 EL 的内置对象
C. EL 有 11 个隐式对象，常用的有 pageScope、reuqestScope、sessionScope、applicationScope
D. EL 和 JSTL 相辅相成，共同丰富 JSP 的功能

三、编程题

1. 使用 JSTL 从 request 作用域中获取元素，并通过标签遍历 List 集合，运行结果如图 7-27 所示。

图 7-27　通过标签遍历 List 集合的运行结果

2. 应用<c:forEach>标签，在页面上打印 10 个"Hello World"，运行结果如图 7-28 所示。

图 7-28　运行结果 1

3. 输入一个年龄值，如果值在 0～6 则显示"儿童"，值在 7～17 则显示"青少年"，值大于等于 18 则显示"长大了"，运行结果如图 7-29 所示。

图 7-29　运行结果 2

4. 创建 List 集合，在该集合中添加元素"aa""bb""cc"，并遍历显示集合元素，运行结果如图 7-30 所示。

图 7-30　运行结果 3

项目八　智慧金融信贷管理系统

☀ 项目要求

由于贷款业务蓬勃发展，为了帮助有关部门更好地了解我国各城市个人信贷的现状，帮助银行等金融机构开发完整的借贷系统，降低信用贷款风险率，我们在此构建了智慧金融信贷管理系统。本项目主要将之前所学知识点结合起来，基于各种需求完整实现智慧金融信贷管理系统。

☀ 项目分析

本项目是基于 MVC 开发模式的金融信贷数据分析平台的开发，整个金融信贷数据分析平台的实现过程是先将准备好的数据存入 MySQL 数据库，使用 MVC 开发模式调取数据库数据，进行前后端交互，最后通过 Echarts 实现数据可视化。该项目分为 6 个任务，即智慧金融信贷管理系统搭建、注册功能实现、登录功能实现、贷款申请功能实现、管理员登录功能实现，以及贷款用户信息查询功能实现。

☀ 项目目标

【知识目标】融会贯通 JDBC 数据库、Servlet、内置对象、EL 表达式等知识。
【能力目标】具备网站综合开发能力。
【素质目标】提升学生发现问题、分析问题、解决问题的能力。

知识导图

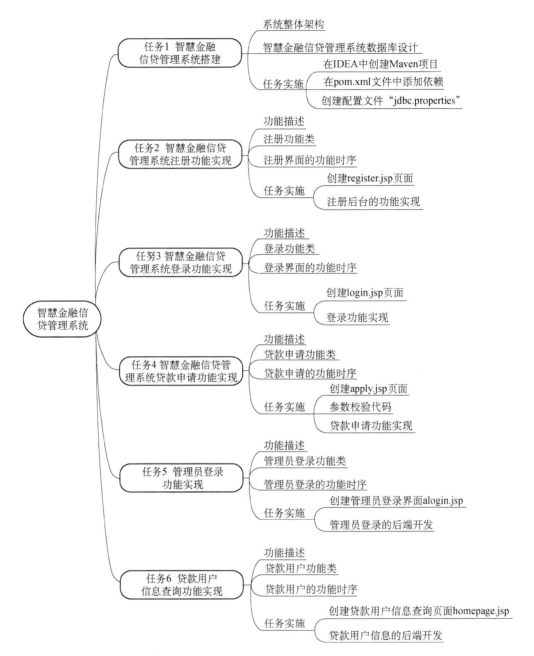

任务 1　智慧金融信贷管理系统搭建

任务演示

教学视频

智慧金融信贷管理系统的主要功能包括注册功能、登录功能、贷款申请功能、管理员登录功能、贷款用户信息查询功能，登录界面、注册界面、贷款申请界面、管理员登录界面、信贷后台系统主页分别如图 8-1 至图 8-5 所示。

图 8-1 登录界面

图 8-2 注册界面

图 8-3 贷款申请界面

图 8-4　管理员登录界面

图 8-5　信贷后台系统主页

本项目介绍的智慧金融信贷管理系统为功能删减版，功能完整版还包括额度分析、违约分析、利率分析、信贷预测、黑名单统计、黑名单信息及审批通过与驳回等功能。功能完整版的代码与功能删减版的代码均在本书附带的电子资源包中，有兴趣的同学可自行下载研究。

 知识准备

1. 系统整体架构

系统整体架构如图 8-6 所示。

项目八 智慧金融信贷管理系统 / 259

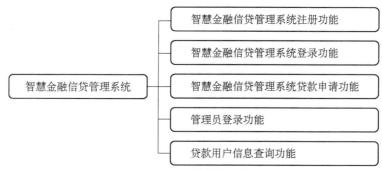

图 8-6 系统整体架构

2. 智慧金融信贷管理系统数据库设计

数据库设计同时为各类用户和各种应用系统提供了信息基础设施和高效的运行环境。数据库是数据共享系统的核心和基础,本系统采用的是关系数据库管理系统 MySQL。创建一个名为"finance_manage"的数据库,该数据库包含 36 个表,本书只节选了部分表进行讲述,在此只需创建用户信息表(user)、贷款申请信息表(apply)、管理员信息表(guser)。

教学视频

(1) 用户信息表

用户信息表(user)用来存储用户的基本信息,该表的结构如表 8-1 所示。

表 8-1 用户信息表的结构

序号	名称	说明	类型	长度	备注
1	id	id	int	11	主键自增
2	username	用户姓名	varchar	255	
3	password	密码	varchar	255	
4	now_time	注册时间	varchar	255	

该表在实现登录功能、注册功能时会用到。

(2) 贷款申请信息表

贷款申请信息表(apply)用来存储贷款申请时所需的相关信息,该表的结构如表 8-2 所示。

表 8-2 贷款申请信息表的结构

序号	名称	说明	类型	长度	备注
1	id	id	int	11	主键自增
2	name	姓名	varchar	255	
3	sex	性别	varchar	255	
4	year	年龄	varchar	255	
5	phone	手机号	varchar	255	
6	body_phone	身份证号	varchar	255	
7	emp_length	工作年限	varchar	255	
8	home_ownership	房屋状态	varchar	255	

（续表）

序号	名称	说明	类型	长度	备注
9	annual_inc	年收入	varchar	255	
10	want_money	贷款金额	varchar	255	
11	now_time	申请时间	varchar	1000	
12	status	审批状态	varchar	255	

该表在实现贷款申请功能时会用到。

（3）管理员信息表

管理员信息表（guser）用于存放管理员信息，该表的结构如表 8-3 所示。

表 8-3　管理员信息表的结构

序号	名称	说明	类型	长度	备注
1	id	id	int	11	主键自增
2	username	管理员姓名	varchar	255	
3	password	密码	varchar	255	

该表在实现后台管理员登录功能时会用到。

任务实施

1. 在 IDEA 中创建 Maven 项目

① 打开 IDEA，单击 "Create New Project" 选项，如图 8-7 所示。

图 8-7　单击 "Create New Project" 选项

② 在打开的 "New Project" 对话框中双击 "Maven" 选项，再选择 JDK（JDK 是开发人员使用的 SDK，因此图 8-8 中所示为 "Project SDK"）版本，建议选择 1.8 版本。最后单击 "Next" 按钮，如图 8-8 所示。

项目八 智慧金融信贷管理系统

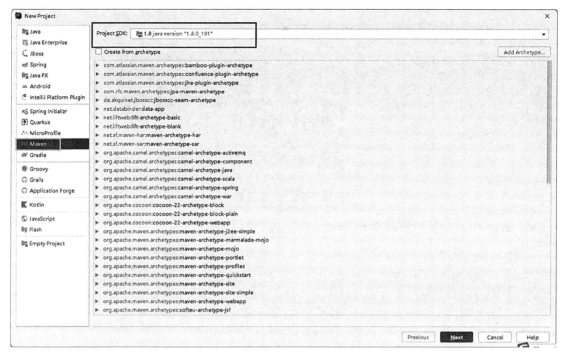

图 8-8 选择 JDK 版本

③ 先输入项目名称"finance_manage",然后单击"Finish"按钮,如图 8-9 所示。至此项目创建完毕。

图 8-9 输入项目名称

④ 在项目中依次单击"src"→"main"→"java"文件夹,然后右键单击"java"文件夹,在弹出的快捷菜单中依次选择"New"→"Package"命令新建包,输入倒序书写的域名包路径"cn.edu.cqcvc",具体步骤如图 8-10 和图 8-11 所示。

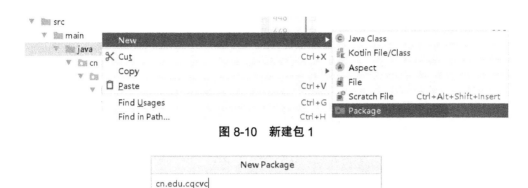

图 8-10　新建包 1

图 8-11　新建包 2

2. 在 pom.xml 文件中添加依赖

项目结构已搭好了，接下来在 pom.xml 文件中添加依赖，添加的文件内容如下。

```xml
<?xml version="1.0" encoding="UTF-8"?>
<project xmlns="http://maven.apache.org/POM/4.0.0"
         xmlns:xsi="http://www.w3.org/2001/XMLSchema-instance"
         xsi:schemaLocation="http://maven.apache.org/POM/4.0.0 http://maven.apache.org/xsd/maven-4.0.0.xsd">
    <modelVersion>4.0.0</modelVersion>
    <groupId>finance_manage</groupId>
    <artifactId>finance_manage</artifactId>
    <version>1.0-SNAPSHOT</version>
    <packaging>war</packaging>
    <name>finance_manage Maven Webapp</name>
    <!-- FIXME change it to the project's website -->
    <url>http://www.example.com</url>
    <properties>
      <project.build.sourceEncoding>UTF-8</project.build.sourceEncoding>
      <maven.compiler.source>1.8</maven.compiler.source>
      <maven.compiler.target>1.8</maven.compiler.target>
      <!-- mysql 版本 -->
      <mysql.version>5.1.6</mysql.version>
    </properties>
    <dependencies>
      <dependency>
        <groupId>org.aspectj</groupId>
        <artifactId>aspectjweaver</artifactId>
        <version>1.8.7</version>
      </dependency>
      <dependency>
        <groupId>taglibs</groupId>
        <artifactId>standard</artifactId>
        <version>1.1.2</version>
      </dependency>
      <!-- mysql -->
```

```xml
<dependency>
  <groupId>mysql</groupId>
  <artifactId>mysql-connector-java</artifactId>
  <version>${mysql.version}</version>
</dependency>
<dependency>
  <groupId>commons-logging</groupId>
  <artifactId>commons-logging</artifactId>
  <version>1.2</version>
</dependency>
<!-- jsp -->
<dependency>
  <groupId>javax.servlet.jsp</groupId>
  <artifactId>jsp-api</artifactId>
  <version>2.2</version>
  <scope>provided</scope>
</dependency>
<!-- servlet -->
<dependency>
  <groupId>javax.servlet</groupId>
  <artifactId>javax.servlet-api</artifactId>
  <version>4.0.1</version>
  <scope>provided</scope>
</dependency>
<!-- fast json -->
<dependency>
  <groupId>com.alibaba</groupId>
  <artifactId>fastjson</artifactId>
  <version>1.2.73</version>
</dependency>
<!-- commons-io -->
<dependency>
  <groupId>commons-io</groupId>
  <artifactId>commons-io</artifactId>
  <version>2.7</version>
</dependency>
<dependency>
  <groupId>jstl</groupId>
  <artifactId>jstl</artifactId>
  <version>1.2</version>
</dependency>
<dependency>
  <groupId>com.github.pagehelper</groupId>
  <artifactId>pagehelper</artifactId>
  <version>5.2.0</version>
</dependency>
</dependencies>
```

```xml
<build>
    <finalName>Loan</finalName>
    <plugins>
      <plugin>
        <groupId>org.apache.tomcat.maven</groupId>
        <artifactId>tomcat7-maven-plugin</artifactId>
        <version>2.2</version>
        <configuration>
          <port>8081</port>
          <path>/</path>
          <uriEncoding>UTF-8</uriEncoding>
        </configuration>
      </plugin>
    </plugins>
    <pluginManagement><!-- lock down plugins versions to avoid using Maven defaults (may be moved to parent pom) -->
      <plugins>
        <plugin>
          <artifactId>maven-clean-plugin</artifactId>
          <version>3.1.0</version>
        </plugin>
        <!-- see http://maven.apache.org/ref/current/maven-core/default-bindings.html#Plugin_bindings_for_war_packaging -->
        <plugin>
          <artifactId>maven-resources-plugin</artifactId>
          <version>3.0.2</version>
        </plugin>
        <plugin>
          <artifactId>maven-compiler-plugin</artifactId>
          <version>3.8.0</version>
        </plugin>
        <plugin>
          <artifactId>maven-surefire-plugin</artifactId>
          <version>2.22.1</version>
        </plugin>
        <plugin>
          <artifactId>maven-war-plugin</artifactId>
          <version>3.2.2</version>
        </plugin>
        <plugin>
          <artifactId>maven-install-plugin</artifactId>
          <version>2.5.2</version>
        </plugin>
        <plugin>
          <artifactId>maven-deploy-plugin</artifactId>
          <version>2.8.2</version>
```

```
          </plugin>
        </plugins>
      </pluginManagement>
    </build>
</project>
```

3. 创建配置文件 "jdbc.properties"

在 resources 文件夹下创建配置文件 "jdbc.properties"，如图 8-12 所示，该配置文件将用于配置数据库连接的四要素，四要素如下所示。

图 8-12　创建配置文件

```
jdbc.driver=com.mysql.jdbc.Driver
jdbc.url=jdbc:mysql://127.0.0.1:3306/project?useUnicode=true&characterEncoding=utf-8&serverTimezone=UTC
jdbc.username=root
jdbc.password=admin
```

注意：若是 8.0 以上版本的数据库，则需要将 "jdbc.driver" 后的配置信息改为 "com.mysql.cj.jdbc.Driver"。至此准备工作完毕。

任务 2　智慧金融信贷管理系统注册功能实现

 任务演示

在一个新用户没有账号时，就需要注册账号，新用户注册成功后会跳转到登录界面。若用户已有账号，可直接登录。注册界面如图 8-2 所示。

教学视频

 知识准备

1. 功能描述

用户进入注册界面后，先输入用户名、密码，然后单击 "注册" 按钮，系统将会核对两次输入的密码，若两次输入的密码相同，则反馈给用户注册成功的提示。注册界面功能描述如图 8-13 所示。

2. 注册功能类

注册功能类如图 8-14 所示。

图 8-13 注册界面功能描述

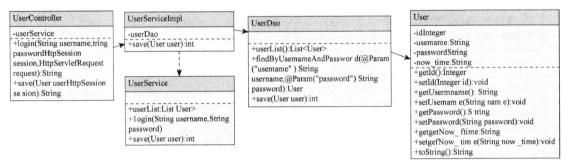

图 8-14 注册功能类

3. 注册界面的功能时序

注册界面的功能时序如图 8-15 所示。

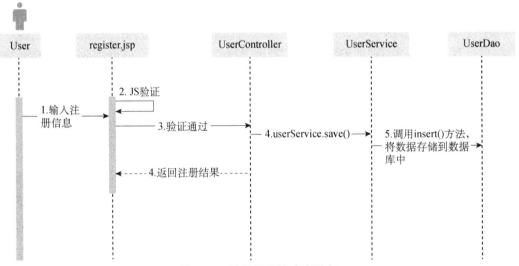

图 8-15 注册界面的功能时序

任务实施

1. 创建 register.jsp 页面

该模块的视图部分是由一个 JSP 页面构成的，这个 JSP 页面是 register.jsp，负责将用户的注册信息提交到 Servlet 控制器 registerServlet 中，并显示注册是否成功，代码如下，效果如图 8-2 所示。

```
<%@ page language="java" contentType="text/html;charset=UTF-8"
```

```jsp
        pageEncoding="UTF-8"%>
<!DOCTYPE html>
<html>
<head>
<meta charset="UTF-8">
<title>智慧金融信贷管理系统</title>
<link href='http://fonts.useso.com/css?family=Open+Sans:400,300,400italic,700' rel='stylesheet' type='text/css'>
        <link href="${pageContext.request.contextPath}/static/assets/bootstrap/css/font-awesome.min.css" rel="stylesheet">
        <link href="${pageContext.request.contextPath}/static/assets/bootstrap/css/bootstrap.min.css" rel="stylesheet">
        <link href="${pageContext.request.contextPath}/static/assets/bootstrap/css/templatemo-style.css" rel="stylesheet">
        <script src="${pageContext.request.contextPath}/static/mjq/js/jquery-3.2.0.min.js"></script>
        <style>
            body {
                background-image:url("/static/assets/img/backgrounds/4.jpg");
            }
        </style>
</head>
<body>
<div class="templatemo-content-widget templatemo-login-widget white-bg">
    <header class="text-center">
        <img src="${pageContext.request.contextPath}/static/wj/picture/518.jpeg" style="width:40px;">
        <h1>信贷数据分析可视化平台</h1>
        <h2>用户注册</h2>
    </header>
    <form id="subform" action="${pageContext.request.contextPath}/user/save" method="post" class="templatemo-login-form" onsubmit="return checkForm(this);">
        <div class="form-group">
            <div class="input-group">
                <div class="input-group-addon"><i class="fa fa-user fa-fw"></i></div>
                <input type="text" class="form-control" name="username" placeholder="用户名">
            </div>
        </div>
        <div class="form-group">
            <div class="input-group">
                <div class="input-group-addon"><i class="fa fa-key fa-fw"></i></div>
                <input id="pwd" type="password" class="form-control" name="password" placeholder="密码">
            </div>
```

```html
            </div>
            <div class="form-group">
                <div class="input-group">
                    <div class="input-group-addon"><i class="fa fa-key fa-fw"></i></div>
                    <input id="rpwd" type="password" class="form-control" name="password1" placeholder="确认密码">
                </div>
            </div>
            <div class="form-group">
                <button id="btn" type="submit" class="templatemo-blue-button width-100">注册</button>
            </div>
            <p style="text-align:center">
                已有账户?<a href="login.jsp">去登录</a>
            </p>
            <p>
                ${sessionScope.add1}
            </p>
        </form>
    </div>
    <script>
        $(function () {
            /*
            * 单击"注册"按钮
            *     判断2次输入的密码是否一致
            *     如果一致,则提交到后台,否则提示错误
            * */
            $('#btn').click(function () {
                //获取2次输入的密码
                var pwd = $('#pwd').val();

                var rpwd=$('#rpwd').val();

                if(pwd===rpwd){
                    //提交到后台
                    $('#subform').submit();
                }else{
                    //提示
                    alert("2次密码输入有误,请重新注册");
                    return false;
                }
            });
        });
    </script>
    <script type="text/javascript">
        // 验证输入不为空的脚本代码
```

```
        function checkForm(form) {
            if(form.username.value === "") {
                alert("用户名不能为空!");
                form.username.focus();
                return false;
            }
            if(form.password.value === "") {
                alert("密码不能为空!");
                form.password.focus();
                return false;
            }
            return true;
        }
    </script>
    </body>
</html>
```

代码解析：在此 JSP 页面中引入了大量 bootstrap 的样式和其他样式，所以代码看上去有些复杂，不过不用担心，核心代码与逻辑位于<form></form>表单标签之间，我们只需要关注<input>标签中的 name 属性。因为在后端的 Servlet 会通过 name 属性的值来获取与其对应的传递值。

上述代码的最后两个<script></script>标签中，有两段校验数据完整性的代码，分别用于校验两次输入的密码是否一致及是否为空值。

2. 注册后台的功能实现

（1）创建 RegisterServlet 类

RegisterServlet 类主要用来接收注册界面发来的 "/user/save" 请求，核心代码如下。

```
package cn.edu.cqcvc.servlet;
import cn.edu.cqcvc.dao.UserDao;
import cn.edu.cqcvc.dao.impl.UserDaoImpl;
import cn.edu.cqcvc.domain.User;
import cn.edu.cqcvc.util.Md5Utils;
import javax.servlet.ServletException;
import javax.servlet.annotation.WebServlet;
import javax.servlet.http.HttpServlet;
import javax.servlet.http.HttpServletRequest;
import javax.servlet.http.HttpServletResponse;
import javax.servlet.http.HttpSession;
import java.io.IOException;
import java.text.SimpleDateFormat;
import java.util.Date;
@WebServlet("/user/save")
public class RegisterServlet extends HttpServlet {
    final private UserDao userDaoImpl = new UserDaoImpl();
```

```java
    @Override
    protected void service(HttpServletRequest req,HttpServletResponse resp) throws
ServletException,IOException {
        //接收请求参数
        String username = req.getParameter("username");
        String password = req.getParameter("password");
        password = Md5Utils.code(password);
        //封装对象
        User user = new User();
        user.setPassword(password);
        user.setUsername(username);
        //设置注册时间
        Date currentTime = new Date();
        SimpleDateFormat formatter = new SimpleDateFormat("yyyy-MM-dd");
        String dateString = formatter.format(currentTime);
        user.setNow_time(dateString);
        //增加
        int acount = userDaoImpl.save(user);
        //获取session
        HttpSession session = req.getSession();
        if (acount == 1) {
            //新增成功,返回登录界面
            session.setAttribute("add","注册成功,请登录!");
            session.setMaxInactiveInterval(3);
            resp.sendRedirect("/login.jsp");
        } else {
            //新增失败,返回注册界面
            session.setAttribute("add1","注册失败,请重新注册!");
            session.setMaxInactiveInterval(3);
            resp.sendRedirect("/register.jsp");
        }
    }
}
```

（2）在 UserDao 接口中添加 save()方法

在创建好的 UserDao 接口中添加 save()方法，代码如下。

```java
package cn.edu.cqcvc.dao;
import cn.edu.cqcvc.domain.User;
public interface UserDao {
    int save(User user);
    User login(User user);
}
```

（3）在 UserDaoImpl 类中添加 save()方法

在创建好的 UserDaoImpl 类中添加 save()方法，实现注册功能，核心代码如下。

```java
@Override
    public int save(User user) {
        //获取链接对象
        Connection connection = null;
        //预编译对象
        PreparedStatement preparedStatement = null;
     //获取结果。只返回在执行 SQL 语句时受影响的数据库的相应表的行数,所以返回一个整型数据
        int ret = 0;
        try {
            //获取链接对象
            connection = JDBCUtils.getConnection();
            //向数据库传递预编译 SQL 语句
            String sql = "INSERT INTO `user`(username,`password`,now_time) VALUES (?,?,?)";
            preparedStatement = connection.prepareStatement(sql);
            preparedStatement.setObject(1,user.getUsername());
            preparedStatement.setObject(2,user.getPassword());
            preparedStatement.setObject(3,user.getNow_time());
            ret = preparedStatement.executeUpdate();
        } catch (Exception e) {
            e.printStackTrace();
        } finally {
            JDBCUtils.release(preparedStatement,connection);
        }
        return ret;
    }
```

至此,注册功能已完成。

任务 3　智慧金融信贷管理系统登录功能实现

 任务演示

在用户进入登录界面并输入用户名和密码后,单击"登录"按钮,待系统验证用户名和密码正确后跳转至系统首页。

本次任务需要开发前端页面来完成用户登录操作。当 Tomcat 成功启动后,在浏览器地址栏输入"http://localhost:8080",就可以访问登录界面。登录界面如图 8-1 所示。

教学视频

 知识准备

1. 功能描述

在用户进入登录界面并输入用户名和密码后,单击"登录"按钮,待系统验证用户名

和密码正确后跳转至系统首页。

2. 登录功能类

登录功能类如图 8-16 所示。

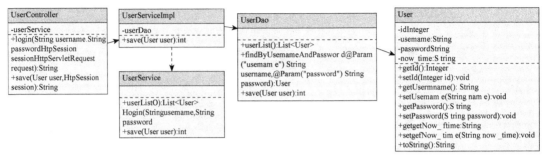

图 8-16　登录功能类

3. 登录界面的功能时序

登录界面的功能时序如图 8-17 所示。

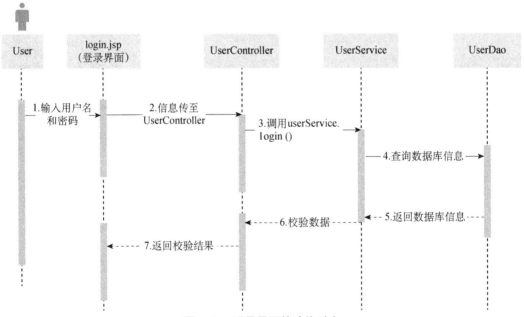

图 8-17　登录界面的功能时序

任务实施

1. 创建 login.jsp 页面

在 webapp 文件夹下创建一个文件 login.jsp，代码如下。

```
<%@ page language="java" contentType="text/html;charset=UTF-8"
    pageEncoding="UTF-8"%>
<!DOCTYPE html>
```

```html
<html>
<head>
<meta charset="UTF-8">
<title>信贷平台登录</title>
    <link href='http://fonts.useso.com/css?family=Open+Sans:400,300,400italic,700' rel='stylesheet' type='text/css'>
    <link href="${pageContext.request.contextPath}/static/assets/bootstrap/css/font-awesome.min.css" rel="stylesheet">
    <link href="${pageContext.request.contextPath}/static/assets/bootstrap/css/bootstrap.min.css" rel="stylesheet">
    <link href="${pageContext.request.contextPath}/static/assets/bootstrap/css/templatemo-style.css" rel="stylesheet">
    <style>
        body{
            background-image:url("/static/assets/img/backgrounds/4.jpg");
        }
    </style>
</head>
<body>
<div class="templatemo-content-widget templatemo-login-widget white-bg">
    <header class="text-center">
        <img src="${pageContext.request.contextPath}/static/wj/picture/518.jpeg" style="width:40px;">
        <h1>信贷数据分析可视化平台</h1>
        <h2>用户登录</h2>
    </header>
    <form action="${pageContext.request.contextPath}/user/login" method="POST" class="templatemo-login-form">
        <div class="form-group">
            <div class="input-group">
                <div class="input-group-addon"><i class="fa fa-user fa-fw"></i></div>
                <input type="text" class="form-control" name="username" placeholder="用户名">
            </div>
        </div>
        <div class="form-group">
            <div class="input-group">
                <div class="input-group-addon"><i class="fa fa-key fa-fw"></i> </div>
                <input type="password" class="form-control" name="password" placeholder="密码">
            </div>
        </div>
        <div class="form-group">
```

```html
            <button type="submit" class="templatemo-blue-button width-100" >登录</button>
        </div>
        <p style="text-align:center">
            还没有账户?<a href="register.jsp">去注册</a>
        </p>
        <p style="text-align:center">
            <a href="guanli.jsp">管理员账户登录</a>
        </p>
        <p>
            ${sessionScope.info}
            ${sessionScope.add}
        </p>
    </form>
</div>
</body>
</html>
```

2. 登录功能实现

（1）创建 LoginServlet 类

接下来开发后台核心控制器，在 cqcvc 包下创建 servlet 包，并且新建 LoginServlet 类，如图 8-18 所示。

图 8-18 新建 LoginServlet 类

该类的作用是接收前端页面 login.jsp 中的用户名和密码，登录时发送的表单请求路径为 "/user/login"，在 LoginServlet 类中接收请求并对参数进行封装，并传递给 Dao 持久层处理，核心代码如下。

```java
package cn.edu.cqcvc.servlet;
import cn.edu.cqcvc.dao.UserDao;
```

```java
import cn.edu.cqcvc.dao.impl.UserDaoImpl;
import cn.edu.cqcvc.domain.User;
import cn.edu.cqcvc.util.Md5Utils;
import javax.servlet.ServletException;
import javax.servlet.annotation.WebServlet;
import javax.servlet.http.HttpServlet;
import javax.servlet.http.HttpServletRequest;
import javax.servlet.http.HttpServletResponse;
import javax.servlet.http.HttpSession;
import java.io.IOException;
import java.text.SimpleDateFormat;
import java.util.Date;
@WebServlet("/user/login")
public class LoginServlet extends HttpServlet {
    final private UserDao userdaoImpl = new UserDaoImpl();
    @Override
    protected void service(HttpServletRequest req,HttpServletResponse resp) throws ServletException,IOException {
        //接收请求参数
        String username = req.getParameter("username");
        String password = req.getParameter("password");
        password= Md5Utils.code(password);
        //封装对象
        User user = new User();
        user.setPassword(password);
        user.setUsername(username);
        //增加
        User loginUser = userdaoImpl.login(user);
        //获取session
        HttpSession session = req.getSession();
        if(loginUser!=null){
            //登录成功 将user存储到session
            session.setAttribute("user",user);
            resp.sendRedirect("/apply.jsp");
        }
        else {
            session.setAttribute("info","账号或密码错误!!!");
            session.setMaxInactiveInterval(3);
            resp.sendRedirect("/login.jsp");
        }
    }
}
```

(2) 在 dao 包中创建 UserDao 接口

接下来写服务层的代码。在 cqcvc 包中创建 dao 包，再在 dao 包中创建 UserDao 接口，如图 8-19 所示。

图 8-19　创建 UserDao 接口

在 servlet 对 dao 包的调用中使用了面向接口编程，接口代码如下。

```
package cn.edu.cqcvc.dao;
import cn.edu.cqcvc.domain.User;
public interface UserDao {
    User login(User user);
}
```

（3）创建实现 UserDao 接口的类

继续在 dao 包中创建 impl 包，再在 impl 包中创建 UserDaoImpl 类，如图 8-20 所示，并在此类中实现 UserDao 接口。

图 8-20　创建 UserDaoImpl 类

UserDaoImpl 类主要用于实现 UserDao 接口中的 login()方法，具体代码如下。

```java
@Override
    public User login(User user) {
        //获取链接对象
        Connection connection = null;
        //预编译对象
        PreparedStatement preparedStatement = null;
        //接收结果集
        ResultSet resultSet = null;
        try {
            //获取链接对象
            connection = JDBCUtils.getConnection();
            //向数据库传递预编译SQL语句
            String sql = "SELECT id,username,`password` FROM `user` WHERE username = ? AND `password` = ?";
            preparedStatement = connection.prepareStatement(sql);
            preparedStatement.setObject(1,user.getUsername());
            preparedStatement.setObject(2,user.getPassword());
            resultSet = preparedStatement.executeQuery();
            if (resultSet.next()) {
                String username = (String) resultSet.getObject("username");
                String password = (String) resultSet.getObject("password");
                User res = new User();
                res.setUsername(username);
                res.setPassword(password);
                return res;
            }
        } catch (Exception e) {
            e.printStackTrace();
        } finally {
            JDBCUtils.release(resultSet,preparedStatement,connection);
        }
        return null;
    }
```

上述代码的编写逻辑是判断通过 JDBC 从数据库查询页面上传来的 username 值和 password 值是否存在，若存在，则说明用户在 login.jsp 页面中输入的用户名和密码是正确的，反之则登录失败。

注意：这里的 JDBC 操作，如加载驱动、获取链接对象、释放资源等，全部被封装为 JDBCUtils 工具类，所以在获取链接对象时，直接调用了该工具类中封装好的 getConnection() 方法。在 finally 代码块中释放资源也是直接调用的 release()方法。JDBCUtils 工具类代码如下。

```java
package cn.edu.cqcvc.util;
import java.io.InputStream;
import java.sql.*;
import java.util.Properties;
/**
```

```java
 * JDBC 工具类
 */
public class JDBCUtils {
    private static Properties p;
    static {
            // 加载注册驱动
            // 该操作在整个程序中只需要执行一次
        try {
            // 配置文件只需要加载一次
            InputStream in = Thread.currentThread().getContextClassLoader().getResourceAsStream("jdbc.properties");
            p = new Properties();
            p.load(in);
            Class.forName(p.getProperty("jdbc.driver"));
        } catch (Exception e) {
            e.printStackTrace();
        }
    }
    /**
     * 获得链接的方法
     */
    public static Connection getConnection() throws Exception {
        return DriverManager.getConnection(
                p.getProperty("jdbc.url"),
                p.getProperty("jdbc.username"),
                p.getProperty("jdbc.password")
        );
    }
    /**
     * 资源释放
     */
    public static void release(Statement statement,Connection connection) {
        if (statement != null) {
            try {
                statement.close();
            } catch (SQLException throwables) {
                throwables.printStackTrace();
            }
        }
        if (connection != null) {
            try {
                connection.close();
            } catch (SQLException throwables) {
                throwables.printStackTrace();
            }
        }
    }
    public static void release(ResultSet resultSet,Statement statement,Connection connection) {
```

```
        if (resultSet != null) {
            try {
                resultSet.close();
            } catch (SQLException throwables) {
                throwables.printStackTrace();
            }
        }
        if (statement != null) {
            try {
                statement.close();
            } catch (SQLException throwables) {
                throwables.printStackTrace();
            }
        }
        if (connection != null) {
            try {
                connection.close();
            } catch (SQLException throwables) {
                throwables.printStackTrace();
            }
        }
    }
}
```

到此，登录功能已完成。但是，若此时直接在数据库的 user 表中添加对应的用户名和密码，那么在项目启动后，即使输入正确的用户名和密码，还是无法正常登录，原因是在注册时，对保存在数据库中的密码的"password"字段进行了 MD5 加密，加密后的秘文保存在数据库中，于是在数据校验时，会对传入的密码进行一次加密，这样才能够和数据库中保存的秘文进行一致性的校验。

（4）创建加密的工具类"Md5Utils"

在 servlet 中，加密的工具类"Md5Utils"的代码如下。

```
package cn.edu.cqcvc.util;
import java.security.MessageDigest;
import java.security.NoSuchAlgorithmException;
public class Md5Utils {
    public static String code(String str){
        try{
            //1.获取 MessageDigest 对象,生成一个 MD5 加密计算摘要
            MessageDigest md = MessageDigest.getInstance("MD5");
            /*
            str.getBytes()
            * 使用平台默认的字符集将此字符串编码为 byte 序列,并将结果存储到一个新的 byte 数组中。
            此方法多用在字节流中,用于将字符串转换为字节
            * */

            // 使用指定的字节数组更新摘要 md
            md.update(str.getBytes());
```

```
        /*
         * digest()最后确定返回md5 hash值,返回值为8位的字符串
         * 因为md5 hash值是16位的hex值,实际上是8位的
         * */
        byte[] byteDigest = md.digest();
        int i;
        StringBuffer buf = new StringBuffer("");
        //遍历byteDigest
        //加密逻辑,可以通过debug自行了解加密逻辑
        for(int offset = 0;offset<byteDigest.length;offset++){
            i = byteDigest[offset];
            if(i < 0)
                i += 256;
            if(i < 16)
                buf.append("0");
            //将8位的字符串转换成16位hex值,用字符串来表示;得到字符串形式的hash值
            buf.append(Integer.toHexString(i));
        }
        return buf.toString();
    }catch (NoSuchAlgorithmException e){
        e.printStackTrace();
        return null;
    }
  }
}
```

在 cqcvc 包下创建新包 util,并且将此工具类与上面提到的 JDBCUtils 工具类全部放在 util 包下,使项目结构更加清晰。

任务 4　智慧金融信贷管理系统贷款申请功能实现

任务演示

本任务是为用户提供贷款申请界面,这个页面分为三个模块:一是"信息填写"模块,用户可以在这个模块填写申请数据,单击"提交"按钮即可申请贷款;二是"贷款流程"模块,这个模块提供了贷款申请的注意事项及问题答疑;三是"联系我们"模块。贷款申请界面如图 8-3 所示。

教学视频

知识准备

1. 功能描述

在贷款申请界面输入姓名、年龄、性别、手机号、身份证号、工作年限、住房类型、年收入数、贷款金额,单击"提交"按钮即可进行贷款申请,如图 8-21 所示。

图 8-21 信贷申请功能描述图

2. 贷款申请功能类

贷款申请功能类如图 8-22 所示。

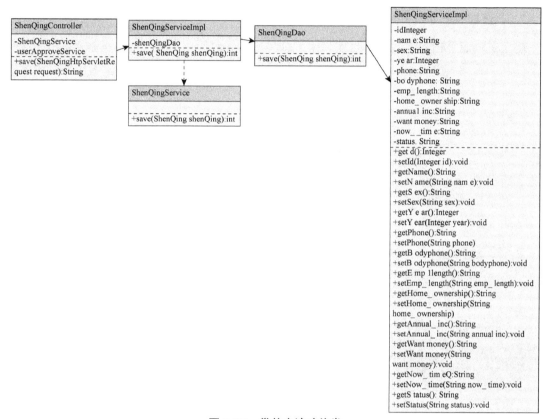

图 8-22 贷款申请功能类

3. 贷款申请的功能时序

贷款申请的功能时序如图 8-23 所示。

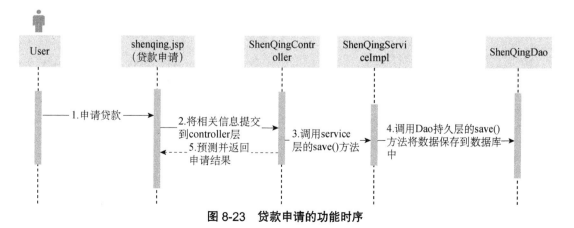

图 8-23 贷款申请的功能时序

任务实施

1. 创建 apply.jsp 页面

在 webapp 文件夹下创建 apply.jsp 页面，此页面的主要功能是填写贷款申请信息，该页面的核心代码可扫描侧方二维码查看。

创建 apply.jsp 页面

2. 参数校验代码

用户在页面上输入参数时，可能会因为种种原因导致输入一些不合规的参数，比如参数为空。开发者应该考虑到这个问题，在参数提交时对相关参数进行校验。参数校验的核心代码如下。

```javascript
<script type="text/javascript">
    // 验证输入参数不为空的脚本代码
    function checkForm(form1) {
        if(form1.name.value === "") {
            alert("请填写真实姓名!!!");
            form1.name.focus();
            return false;
        }
        if(form1.year.value === "") {
            alert("请填写年龄!!!");
            form1.year.focus();
            return false;
        }
        if(form1.sex.value === "") {
            alert("请填写性别!!!");
            form1.sex.focus();
            return false;
        }
        if(form1.phone.value === "") {
            alert("请填写手机号!!!");
            form1.phone.focus();
            return false;
        }
        if(form1.bodyphone.value === "") {
            alert("请填写身份证号!!!");
            form1.bodyphone.focus();
            return false;
        }
        if(form1.emp_length.value === "") {
            alert("请填写工作年限!!!");
            form1.emp_length.focus();
            return false;
        }
        if(form1.home_ownership.value === "") {
```

```
            alert("请填写住房类型!!!");
            form1.home_ownership.focus();
            return false;
        }
        if(form1.annual_inc.value === "") {
            alert("请填写年收入数额!!!");
            form1.annual_inc.focus();
            return false;
        }
        if(form1.want_money.value === "") {
            alert("请填写贷款金额!!!");
            form1.want_money.focus();
            return false;
        }
        if(form1.year.value<18) {
            alert("未成年人不允许贷款!!!");
            form1.year.focus();
            return false;
        }
        if(form1.want_money.value>500000) {
            alert("最高贷款金额为 50 万!!!");
            form1.want_money.focus();
            return false;
        }
        return true;
    }
</script>
```

3. 贷款申请功能实现

① 在项目的 servlet 包中创建 ApplyServlet 类，如图 8-24 所示。

图 8-24 创建 ApplyServlet 类

创建的这个类主要用于接收 apply.jsp 页面发出的请求，并获取参数，核心代码如下。

```java
package cn.edu.cqcvc.servlet;
import cn.edu.cqcvc.dao.ApplyDao;
import cn.edu.cqcvc.dao.impl.ApplyDaoImpl;
import cn.edu.cqcvc.domain.ApplyUser;
import javax.servlet.ServletException;
import javax.servlet.annotation.WebServlet;
import javax.servlet.http.HttpServlet;
import javax.servlet.http.HttpServletRequest;
import javax.servlet.http.HttpServletResponse;
import javax.servlet.http.HttpSession;
import java.io.IOException;
import java.text.SimpleDateFormat;
import java.util.Date;
@WebServlet("/applyUser")
public class ApplyServlet extends HttpServlet {
    final private ApplyDao applyDaoImpl = new ApplyDaoImpl();
    @Override
    protected void service(HttpServletRequest req,HttpServletResponse resp) throws ServletException,IOException {
        //接收请求参数
        String name = req.getParameter("name");
        String year = req.getParameter("year");
        String sex = req.getParameter("sex");
        String phone = req.getParameter("phone");
        String bodyPhone = req.getParameter("bodyphone");
        String empLength = req.getParameter("emp_length");
        String homeOwnership = req.getParameter("home_ownership");
        String annualInc = req.getParameter("annual_inc");
        String wantMoney = req.getParameter("want_money");
        //封装对象
        ApplyUser applyUser = new ApplyUser();
        applyUser.setName(name);
        applyUser.setYear(Integer.parseInt(year));
        applyUser.setSex(sex);
        applyUser.setPhone(phone);
        applyUser.setBodyPhone(bodyPhone);
        applyUser.setEmpLength(empLength);
        applyUser.setHomeOwnership(homeOwnership);
        applyUser.setAnnualInc(annualInc);
        applyUser.setWantMoney(wantMoney);
        applyUser.setStatus("待审核");
        //设置申请时间
        Date currentTime = new Date();
        SimpleDateFormat formatter = new SimpleDateFormat("yyyy-MM-dd");
        String dateString = formatter.format(currentTime);
        applyUser.setNow_time(dateString);
        //获取session
        HttpSession session = req.getSession();
        //调用保存方法
        int account = applyDaoImpl.save(applyUser);
        if (account == 1) {
```

```
            //新增成功
            session.setAttribute("apply","提交成功,请等待审批结果!");
            session.setMaxInactiveInterval(5);
            resp.sendRedirect("/apply.jsp");
        } else {
            //新增失败
            session.setAttribute("apply1","提交失败,请核实信息后再次申请!");
            session.setMaxInactiveInterval(3);
            resp.sendRedirect("/apply.jsp");
        }
    }
}
```

注意：功能完整版还有审核功能，但审核功能暂不作为这里的学习内容，所以这部分代码已省略。

② 在 dao 包中创建 ApplyDao 接口，如图 8-25 所示。

图 8-25　创建 ApplyDao 接口

此接口定义了一个 save()方法，用于接收 servlet 传来的参数，核心代码如下。

```
package cn.edu.cqcvc.dao;
import cn.edu.cqcvc.domain.ApplyUser;
import java.util.List;
public interface ApplyDao {
    int save(ApplyUser applyUser);
}
```

③ 在 impl 包中创建 ApplyDaoImpl 类，如图 8-26 所示。

此类用于实现 ApplyDao 接口中的 save()方法，可用来完成申请贷款信息的具体保存逻辑，核心代码如下。

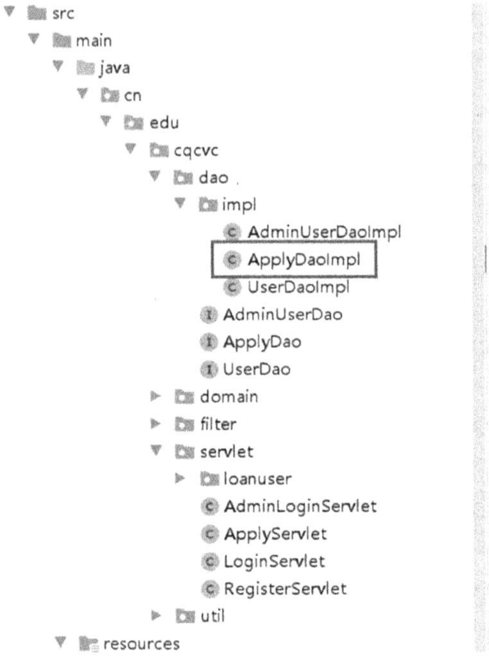

图 8-26 创建 ApplyDaoImpl 类

```java
package cn.edu.cqcvc.dao.impl;
import cn.edu.cqcvc.dao.ApplyDao;
import cn.edu.cqcvc.domain.ApplyUser;
import cn.edu.cqcvc.domain.User;
import cn.edu.cqcvc.util.JDBCUtils;
import java.sql.Connection;
import java.sql.PreparedStatement;
import java.sql.ResultSet;
import java.util.ArrayList;
import java.util.List;
public class ApplyDaoImpl implements ApplyDao {
    @Override
    public int save(ApplyUser applyUser) {
        //获取链接对象
        Connection connection = null;
        //预编译对象
        PreparedStatement preparedStatement = null;
        //获取结果,只返回在执行 SQL 语句时受影响的数据库的相应表的行数,返回一个整型数据。
        int ret = 0;
        try {
            //获取链接对象
            connection = JDBCUtils.getConnection();
            //向数据库传递预编译的 SQL 语句
            String sql = "INSERT INTO apply('name',sex,'year',phone,body_phone,emp_length,home_ownership,annual_inc,want_money,now_time,'status') VALUES(?,?,?,?,?,?,?,?,?,?,?)";
```

```
                preparedStatement = connection.prepareStatement(sql);
                preparedStatement.setObject(1,applyUser.getName());
                preparedStatement.setObject(2,applyUser.getSex());
                preparedStatement.setObject(3,applyUser.getYear());
                preparedStatement.setObject(4,applyUser.getPhone());
                preparedStatement.setObject(5,applyUser.getBodyPhone());
                preparedStatement.setObject(6,applyUser.getEmpLength());
                preparedStatement.setObject(7,applyUser.getHomeOwnership());
                preparedStatement.setObject(8,applyUser.getAnnualInc());
                preparedStatement.setObject(9,applyUser.getWantMoney());
                preparedStatement.setObject(10,applyUser.getNow_time());
//此处给审核状态字段 status 赋值。由于本任务没有实现审核功能,所以此处直接赋值"已通过"
                preparedStatement.setObject(11,"已通过");
                ret = preparedStatement.executeUpdate();
        } catch (Exception e) {
            e.printStackTrace();
        } finally {
            JDBCUtils.release(preparedStatement,connection);
        }
        return ret;
    }
}
```

至此,贷款申请功能已完成。

任务 5　管理员登录功能实现

教学视频

任务演示

智慧金融信贷管理系统分为前台系统与后台系统(即信贷后台系统),前台系统主要有贷款申请功能,在用户申请贷款完成后,管理员可以登录后台系统找到与之对应的申请记录。

管理员有单独的登录界面地址,在浏览器中输入管理员的登录界面地址后,可以进入信贷后台监控平台的管理员登录界面,输入管理员名和密码后,即可正常登录。管理员登录界面如图 8-4 所示。

知识准备

1. 功能描述

管理员输入管理员名和密码,单击"登录"按钮,即可进入信贷后台系统主页,管理员登录流程如图 8-27 所示。

2. 管理员登录功能类

管理员登录功能类如图 8-28 所示。

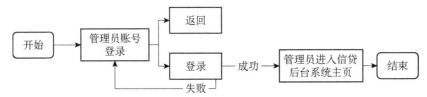

图 8-27　管理员登录流程

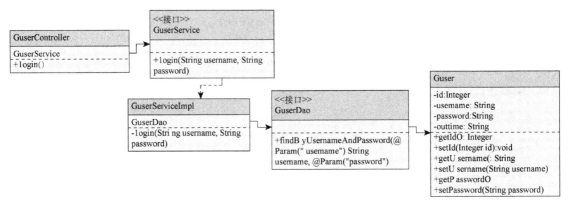

图 8-28　管理员登录功能类

3. 管理员登录的功能时序

管理员登录的功能时序如图 8-29 所示。

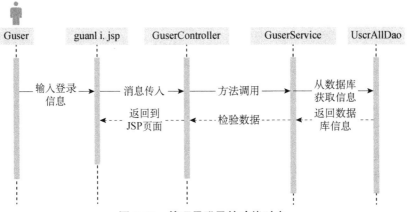

图 8-29　管理员登录的功能时序

任务实施

1. 创建管理员登录界面 alogin.jsp

① 和 login.jsp 一样，先在 webapp 文件夹下创建 alogin.jsp，此界面的功能与 login.jsp 大致相同，其代码如下。

```
<%@ page contentType="text/html;charset=UTF-8" language="java" %>
<!DOCTYPE html>
<html lang="en">
```

```html
<head>
    <meta charset="utf-8">
    <meta http-equiv="X-UA-Compatible" content="IE=edge">
    <meta name="viewport" content="width=device-width,initial-scale=1">
    <title>信贷后台监控平台登录</title>
    <meta name="description" content="">
    <meta name="author" content="templatemo">
  <link href='http://fonts.useso.com/css?family=Open+Sans:400,300,400italic,700' rel='stylesheet' type='text/css'>
    <link    href="${pageContext.request.contextPath}/static/assets/bootstrap/css/font-awesome.min.css" rel="stylesheet">
    <link    href="${pageContext.request.contextPath}/static/assets/bootstrap/css/bootstrap.min.css" rel="stylesheet">
    <link    href="${pageContext.request.contextPath}/static/assets/bootstrap/css/templatemo-style.css" rel="stylesheet">
    <style>
        body {
            background-image:url("/static/assets/img/backgrounds/4.jpg");
        }
    </style>
</head>
<body>
<div class="templatemo-content-widget templatemo-login-widget white-bg">
    <header class="text-center">
   <img src="${pageContext.request.contextPath}/static/wj/picture/518.jpeg" style="width:40px;">
        <h1>信贷后台监控平台</h1>
        <h2>管理员登录</h2>
    </header>
    <form action="${pageContext.request.contextPath}/admin/login" method="post" class="templatemo-login-form">
        <div class="form-group">
            <div class="input-group">
                <div class="input-group-addon"><i class="fa fa-user fa-fw"></i></div>
                <input type="text" class="form-control" name="username" placeholder="管理员名">
            </div>
        </div>
        <div class="form-group">
            <div class="input-group">
                <div class="input-group-addon"><i class="fa fa-key fa-fw"></i> </div>
                <input type="password" class="form-control" name="password" placeholder=
```

```
"密码">
            </div>
        </div>
        <div class="form-group">
            <button type="submit" class="templatemo-blue-button width-100">登录</button>
        </div>
        <p style="text-align:center">
            <a href="login.jsp">返回</a>
        </p>
        <p>
            ${sessionScope.info1}
        </p>
    </form>
</div>
</body>
</html>
```

2. 管理员登录的后端开发

① 在 servlet 包下创建 AdminLoginServlet 类，如图 8-30 所示。

图 8-30　创建 AdminLoginServlet 类

和普通用户登录一样，管理员登录也需要使用 servlet 接收页面传入的"username"和"password"，核心代码如下。

```
package cn.edu.cqcvc.servlet;
import cn.edu.cqcvc.dao.AdminUserDao;
import cn.edu.cqcvc.dao.impl.AdminUserDaoImpl;
import cn.edu.cqcvc.domain.AdminUser;
```

```java
import cn.edu.cqcvc.util.Md5Utils;
import javax.servlet.ServletException;
import javax.servlet.annotation.WebServlet;
import javax.servlet.http.HttpServlet;
import javax.servlet.http.HttpServletRequest;
import javax.servlet.http.HttpServletResponse;
import javax.servlet.http.HttpSession;
import java.io.IOException;
@WebServlet("/admin/login")
public class AdminLoginServlet extends HttpServlet {
    final private AdminUserDao userDaoImpl = new AdminUserDaoImpl();
    @Override
    protected void service(HttpServletRequest req,HttpServletResponse resp) throws ServletException,IOException {
        //接收请求参数
        String username = req.getParameter("username");
        String password = req.getParameter("password");
        password= Md5Utils.code(password);
        //封装对象
        AdminUser user = new AdminUser();
        user.setPassword(password);
        user.setUsername(username);
        //增加
        AdminUser loginUser = userDaoImpl.login(user);
        //获取session
        HttpSession session = req.getSession();
        if(loginUser!=null){
            //登录成功,将user存储到session中
            session.setAttribute("guser",loginUser);
            resp.sendRedirect("/pages/first.jsp");
        }
        else {
            session.setAttribute("info1","账号或密码错误!!!");
            session.setMaxInactiveInterval(3);
            resp.sendRedirect("/alogin.jsp");
        }
    }
}
```

和普通用户登录的不同之处在于,管理员登录成功后,登录信息会被保存在 session 中。

注意:session 是保存在服务器中的内存空间,一般用于保存用户的会话信息。

这样做的好处在于,如果当前用户在接下来的某一段时间内再次访问该页面时,系统就可以识别到 session 中已有的这个用户的信息,不必再次登录。

② 在 dao 包下创建 AdminUserDao 接口,如图 8-31 所示。

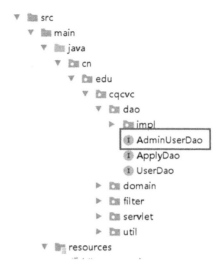

图 8-31 创建 AdminUserDao 接口

此接口与普通用户登录的接口差不多，源代码如下。

```
package cn.edu.cqcvc.dao;
import cn.edu.cqcvc.domain.AdminUser;
public interface AdminUserDao {
    AdminUser login(AdminUser user);
}
```

③ 在 impl 包下创建 AdminUserDaoImpl 类，如图 8-32 所示。

图 8-32 创建 AdminUserDaoImpl 类

该类的核心代码如下。

```
package cn.edu.cqcvc.dao.impl;
import cn.edu.cqcvc.dao.AdminUserDao;
import cn.edu.cqcvc.domain.AdminUser;
import cn.edu.cqcvc.util.JDBCUtils;
import java.sql.Connection;
import java.sql.PreparedStatement;
```

```java
import java.sql.ResultSet;
public class AdminUserDaoImpl implements AdminUserDao {
    @Override
    public AdminUser login(AdminUser user) {
        //获取链接对象
        Connection connection = null;
        //预编译对象
        PreparedStatement preparedStatement = null;
        //接收结果集
        ResultSet resultSet = null;
        try {
            //获取链接对象
            connection = JDBCUtils.getConnection();
            //向数据库传递预编译的SQL语句
            String sql = "SELECT id,username,`password` FROM `guser` WHERE username = ? AND `password` = ?";
            preparedStatement = connection.prepareStatement(sql);
            preparedStatement.setObject(1,user.getUsername());
            preparedStatement.setObject(2,user.getPassword());
            resultSet = preparedStatement.executeQuery();
            if (resultSet.next()) {
                String username = (String) resultSet.getObject("username");
                String password = (String) resultSet.getObject("password");
                AdminUser res = new AdminUser();
                res.setUsername(username);
                res.setPassword(password);
                return res;
            }
        } catch (Exception e) {
            e.printStackTrace();
        } finally {
            JDBCUtils.release(resultSet,preparedStatement,connection);
        }
        return null;
    }
}
```

至此，管理员登录功能已完成。

任务6 贷款用户信息查询功能实现

教学视频

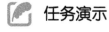 任务演示

在管理员登录信贷后台系统后，就会进入信贷后台系统主页，如图8-5所示。

对于真实的线上项目，其信贷后台系统主页中的"总用户数""总贷款人数""日注册数""日贷款数"是从数据库中查询的实时数据；"住房状态""年平均利率""每年贷款人数""平均单笔贷款额度"四张图表是使用 Echarts 导入真实数据创建的图表。本项目为实训项目，故这些数据都是固定数据。若想要研究上述知识点，请自行参考在线教学资源中的完整源代码。

信贷后台系统主页采用 Layui+Bootstrap 框架实现，这里不做重点研究，感兴趣的同学自行研究，可直接从在线教学资源的源代码中复制。

注意：信贷后台系统主页发送了许多数据查询的请求，但有些请求对应的查询功能并没有实现，同学们可根据需求自行实现。

在信贷后台系统主页显示出来后，单击左侧菜单栏的"贷款用户信息"选项，即可进入贷款用户信息功能页面。

 知识准备

1. 功能描述

管理员进入信贷后台系统后，单击"贷款用户信息"选项，即可对用户的姓名、手机号、贷款状态和贷款城市进行查询，也可以新增数据和删除数据，还可以展示"用户数""坏账数""完成还款""贷款中"数据，以及周期为 36 和 60 的人数。贷款用户信息查询流程如图 8-33 所示。

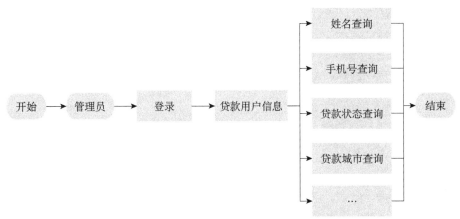

图 8-33　贷款用户信息查询流程

2. 贷款用户功能类

贷款用户功能类如图 8-34 所示。

3. 贷款用户的功能时序

贷款用户时序图如图 8-35 所示。
贷款用户信息选项如图 8-36 所示。
贷款用户信息页面如图 8-37 所示。

基于本实训项目的整体难度规划，以及同学们目前所掌握的知识点，本任务主要完成贷款用户信息的查询功能，其他功能可根据兴趣自行完成。

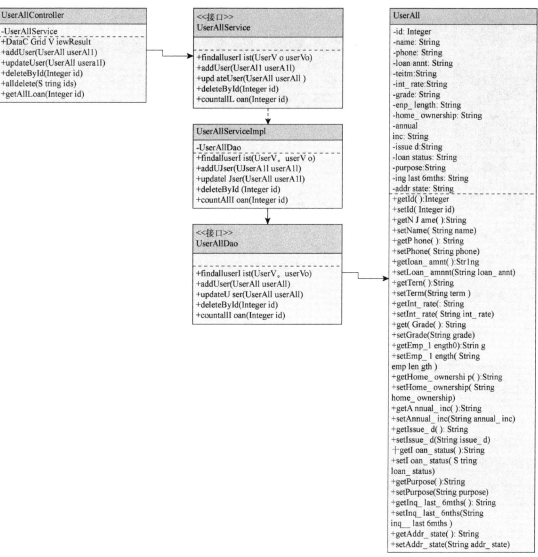

图 8-34 贷款用户功能类

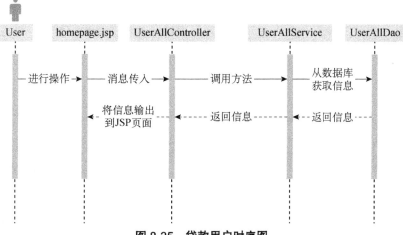

图 8-35 贷款用户时序图

图 8-36 贷款用户
信息选项

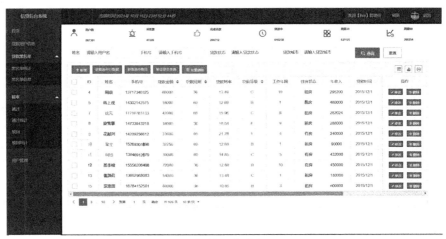

图 8-37 贷款用户信息页面

任务实施

first.jsp 文件可直接从在线教学资源中资源包中复制，因为没有真实数据的支持，所以只用作展示。

1. 创建贷款用户信息查询页面 homepage.jsp

此页面依旧采用 Layui+Bootstrap 框架实现，使用 Layui 框架发送 AJAX 请求，获取返回的 JSON 数据，并且将 JSON 数据展示在页面对应的位置。

在项目路径"webapp\pages"下创建 homepage.jsp 文件，参考代码可扫描侧方二维码查看。

创建贷款用户
信息查询页面
homepage.jsp

2. 贷款用户信息的后端开发

后端开发的主要内容是提供处理贷款用户信息的增、删、改、查方法。在 servlet 包下创建新包 loanuser，并在此包下创建 UserAllServlet 类、UserAddServlet 类、UserDeleteAllServlet 类、UserDeleteServlet 类、UserUpdateServlet 类，用于接收贷款用户信息页面发来的各种请求，如增、删、改、查贷款用户信息等。并将各个请求映射到各自对应的 servlet 即可。

（1）创建 UserAllServlet 类

代码如下：

```java
package cn.edu.cqcvc.servlet.loanuser;
import cn.edu.cqcvc.dao.ApplyDao;
import cn.edu.cqcvc.dao.impl.ApplyDaoImpl;
import cn.edu.cqcvc.domain.ApplyUser;
import cn.edu.cqcvc.util.DataGridViewResult;
import com.alibaba.fastjson.JSON;
import com.github.pagehelper.PageHelper;
import com.github.pagehelper.PageInfo;
```

```java
import javax.servlet.ServletException;
import javax.servlet.annotation.WebServlet;
import javax.servlet.http.HttpServlet;
import javax.servlet.http.HttpServletRequest;
import javax.servlet.http.HttpServletResponse;
import java.io.IOException;
import java.io.PrintWriter;
import java.util.List;
@WebServlet("/userall/find")
public class UserAllServlet extends HttpServlet {
    final private ApplyDao loanUserDaoImpl = new ApplyDaoImpl();
    @Override
    protected void service(HttpServletRequest req,HttpServletResponse resp) throws ServletException,IOException {
        //接收分页请求参数
        //页数
        String page = req.getParameter("page");
        //设置每页显示多少条数据
        String limit = req.getParameter("limit");
        PageHelper.startPage(Integer.parseInt(page),Integer.parseInt(limit));
        //查询所有贷款用户信息
        List<ApplyUser> loginUser = loanUserDaoImpl.selectAll();
        PageInfo<ApplyUser> pageInfo = new PageInfo<>(loginUser);
        DataGridViewResult dataGridViewResult = new DataGridViewResult(pageInfo.getTotal(),pageInfo.getList());
        String jsonStr = JSON.toJSONString(dataGridViewResult);
        PrintWriter writer = resp.getWriter();
        writer.write(jsonStr);
    }
}
```

（2）创建 UserAddServlet 类

代码如下：

```java
package cn.edu.cqcvc.servlet.loanuser;
import cn.edu.cqcvc.dao.ApplyDao;
import cn.edu.cqcvc.dao.impl.ApplyDaoImpl;
import cn.edu.cqcvc.domain.ApplyUser;
import cn.edu.cqcvc.util.DataGridViewResult;
import com.alibaba.fastjson.JSON;
import com.github.pagehelper.PageHelper;
import com.github.pagehelper.PageInfo;
import javax.servlet.ServletException;
import javax.servlet.annotation.WebServlet;
import javax.servlet.http.HttpServlet;
import javax.servlet.http.HttpServletRequest;
import javax.servlet.http.HttpServletResponse;
import java.io.IOException;
```

```java
import java.io.PrintWriter;
import java.util.List;
@WebServlet("/userall/find")
public class UserAllServlet extends HttpServlet {
    final private ApplyDao loanUserDaoImpl = new ApplyDaoImpl();
    @Override
    protected void service(HttpServletRequest req,HttpServletResponse resp) throws ServletException,IOException {
        //接收分页请求参数
        //页数
        String page = req.getParameter("page");
        //设置每页显示多少条数据
        String limit = req.getParameter("limit");
        PageHelper.startPage(Integer.parseInt(page),Integer.parseInt(limit));
        //查询所有贷款用户信息
        List<ApplyUser> loginUser = loanUserDaoImpl.selectAll();
        PageInfo<ApplyUser> pageInfo = new PageInfo<>(loginUser);
        DataGridViewResult dataGridViewResult = new DataGridViewResult(pageInfo.getTotal(),pageInfo.getList());
        String jsonStr = JSON.toJSONString(dataGridViewResult);
        PrintWriter writer = resp.getWriter();
        writer.write(jsonStr);
    }
}
```

（3）创建 UserDeleteAllServlet 类

代码如下：

```java
package cn.edu.cqcvc.servlet.loanuser;
import cn.edu.cqcvc.dao.ApplyDao;
import cn.edu.cqcvc.dao.impl.ApplyDaoImpl;
import cn.edu.cqcvc.domain.ApplyUser;
import cn.edu.cqcvc.util.DataGridViewResult;
import com.alibaba.fastjson.JSON;
import com.github.pagehelper.PageHelper;
import com.github.pagehelper.PageInfo;
import javax.servlet.ServletException;
import javax.servlet.annotation.WebServlet;
import javax.servlet.http.HttpServlet;
import javax.servlet.http.HttpServletRequest;
import javax.servlet.http.HttpServletResponse;
import java.io.IOException;
import java.io.PrintWriter;
import java.util.HashMap;
import java.util.List;
import java.util.Map;
@WebServlet("/userall/alldelete")
public class UserDeleteAllServlet extends HttpServlet {
```

```java
    final private ApplyDao loanUserDaoImpl = new ApplyDaoImpl();
    @Override
    protected void service(HttpServletRequest req,HttpServletResponse resp) throws ServletException,IOException {
        int count = 0;
        //接收请求参数
        String ids = req.getParameter("ids");
        String[] idsStr = ids.split(",");
        //封装返回的Map参数
        Map<String,Object> map = new HashMap<>();
        for (String s:idsStr) {
            count = loanUserDaoImpl.deleteById(Integer.valueOf(s));
            if (count > 0) {
                map.put("success",true);
                map.put("message","删除成功");
            }
        }
        if (count <= 0) {
            map.put("success",false);
            map.put("message","删除失败");
        }
        String jsonStr = JSON.toJSONString(map);
        PrintWriter writer = resp.getWriter();
        writer.write(jsonStr);
    }
}
```

(4) 创建 UserDeleteServlet 类

代码如下：

```java
package cn.edu.cqcvc.servlet.loanuser;
import cn.edu.cqcvc.dao.ApplyDao;
import cn.edu.cqcvc.dao.impl.ApplyDaoImpl;
import com.alibaba.fastjson.JSON;

import javax.servlet.ServletException;
import javax.servlet.annotation.WebServlet;
import javax.servlet.http.HttpServlet;
import javax.servlet.http.HttpServletRequest;
import javax.servlet.http.HttpServletResponse;
import java.io.IOException;
import java.io.PrintWriter;
import java.util.HashMap;
import java.util.Map;
@WebServlet("/userall/deleteById")
public class UserDeleteServlet extends HttpServlet {
    final private ApplyDao loanUserDaoImpl = new ApplyDaoImpl();
    @Override
```

```java
    protected void service(HttpServletRequest req,HttpServletResponse resp) throws
ServletException,IOException {
        int count = 0;
        //接收请求参数
        String id = req.getParameter("id");
        count = loanUserDaoImpl.deleteById(Integer.valueOf(id));
        //封装返回的 Map 参数
        Map<String,Object> map = new HashMap<>();
        if (count > 0) {
           map.put("success",true);
           map.put("message","删除成功");
        } else {
           map.put("success",false);
           map.put("message","删除失败");
        }
        String jsonStr = JSON.toJSONString(map);
        PrintWriter writer = resp.getWriter();
        writer.write(jsonStr);
    }
}
```

(5) 创建 UserUpdateServlet 类

代码如下：

```java
package cn.edu.cqcvc.servlet.loanuser;
import cn.edu.cqcvc.dao.ApplyDao;
import cn.edu.cqcvc.dao.impl.ApplyDaoImpl;
import cn.edu.cqcvc.domain.ApplyUser;
import com.alibaba.fastjson.JSON;
import javax.servlet.ServletException;
import javax.servlet.annotation.WebServlet;
import javax.servlet.http.HttpServlet;
import javax.servlet.http.HttpServletRequest;
import javax.servlet.http.HttpServletResponse;
import java.io.IOException;
import java.io.PrintWriter;
import java.util.HashMap;
import java.util.Map;
@WebServlet("/userall/update")
public class UserUpdateServlet extends HttpServlet {
    final private ApplyDao loanUserDaoImpl = new ApplyDaoImpl();
    @Override
    protected void service(HttpServletRequest req,HttpServletResponse resp) throws
ServletException,IOException {
        //接收请求参数
        String id = req.getParameter("id");
        String name = req.getParameter("name");
        String year = req.getParameter("year");
```

```java
        String sex = req.getParameter("sex");
        String phone = req.getParameter("phone");
        String bodyPhone = req.getParameter("bodyPhone");
        String empLength = req.getParameter("empLength");
        String homeOwnership = req.getParameter("homeOwnership");
        String annualInc = req.getParameter("annualInc");
        String wantMoney = req.getParameter("wantMoney");
        //封装对象
        ApplyUser applyUser = new ApplyUser();
        applyUser.setId(Integer.parseInt(id));
        applyUser.setName(name);
        applyUser.setYear(Integer.parseInt(year));
        applyUser.setSex(sex);
        applyUser.setPhone(phone);
        applyUser.setBodyPhone(bodyPhone);
        applyUser.setEmpLength(empLength);
        applyUser.setHomeOwnership(homeOwnership);
        applyUser.setAnnualInc(annualInc);
        applyUser.setWantMoney(wantMoney);
        applyUser.setStatus("待审核");
        //调用保存方法
        int count = loanUserDaoImpl.update(applyUser);
        //封装返回的 Map 参数
        Map<String,Object> map = new HashMap<>();
        if (count > 0) {
            map.put("success",true);
            map.put("message","修改成功");
        } else {
            map.put("success",false);
            map.put("message","修改失败");
        }
        String jsonStr = JSON.toJSONString(map);
        PrintWriter writer = resp.getWriter();
        writer.write(jsonStr);
    }
}
```

注意：

① UserAllServlet 类中的 PageHelper 类用于完成分页操作，对于数据量较多的查询，分页是必不可少的。本项目未实现分页功能，有兴趣的同学可自行实现。

② 由于贷款用户信息查询页面是通过 Layui 框架实现列表展示的，所以发送的都是 AJAX 请求，需要返回 JSON 数据，而不是请求转发或者重定向到某个页面，这样 Layui 框架的表单组件才能将响应数据展示在页面上。

在创建好的 ApplyDao 接口中实现增、删、改、查方法，代码如下。

```java
package cn.edu.cqcvc.dao;
import cn.edu.cqcvc.domain.ApplyUser;
```

```
import java.util.List;
public interface ApplyDao {
    int save(ApplyUser applyUser);
    List<ApplyUser> selectAll();
    int deleteById(Integer id);
    int update(ApplyUser applyUser);
}
```

（6）实现 ApplyDao 接口

在创建好的 ApplyDaoImpl 类中实现 ApplyDao 接口的增、删、改、查方法，参考代码可扫描侧方二维码查看。

实现 ApplyDao 接口

至此，贷款用户信息查询功能已实现。鉴于本项目为综合实训项目，无课后练习，因此请同学们认真完成项目任务。

本书的全部知识点几乎都应用到了本项目中，所以本项目拥有极强的综合性，可以考验同学们对知识点的掌握程度。通过对本项目的学习，我们了解了一个完整项目的开发流程，打下了良好的工作基础。